One Hundred Prisoners and a Light Bulb

Hans van Ditmarsch • Barteld Kooi

One Hundred Prisoners and a Light Bulb

Illustrations by Elancheziyan

Springer

Copernicus Books is a brand of Springer

Hans van Ditmarsch
LORIA, CNRS
Université de Lorraine
Vandoeuvre-lès-Nancy
France

Barteld Kooi
Faculty of Philosophy
University of Groningen
Groningen
The Netherlands

Illustrations (c)2015 by Hans van Ditmarsch and Barteld Kooi

ISBN 978-3-319-16693-3   ISBN 978-3-319-16694-0 (eBook)
DOI 10.1007/978-3-319-16694-0

Library of Congress Control Number: 2015933648

Mathematics Subject Classification (2010): 00A08

Honderd gevangenen en een gloeilamp
Original Dutch edition published by (c) Epsilon Uitgaven, Amsterdam, 2013

Springer Cham Heidelberg New York Dordrecht London

Printed on acid-free paper

Copernicus is a brand of Springer

Springer International Publishing AG Switzerland is part of Springer Science+Business Media (www.springer.com)

Preface

This puzzlebook presents 11 different puzzles about knowledge and ignorance. Each puzzle is treated in depth in a separate chapter, and each chapter also contains additional puzzles for which the answers can be found at the back of the book. A constant theme in these puzzles is that the persons involved make announcements about what they know and do not know, and then later appear to contradict themselves. Such knowledge puzzles have played an important role in the development of an area known as dynamic epistemic logic. A separate stand-alone chapter gives an introduction to dynamic epistemic logic.

The illustrations for this book were made by Elancheziyan. Elancheziyan is a Tamil speaking Indian illustrator living in Chennai. Hans has an associate position at the Institute of Mathematical Sciences (IMSc) in Chennai, India. By the intermediation of his IMSc host Ramanujam, and the kind assistance of Shubashree Desikan, who acted as a Tamil-English interpreter, he got in contact with Elancheziyan. How the illustrations to each chapter came about is story in itself, and we are very grateful for Elancheziyan's essential part in this joint enterprise.

We wish to thank Paul Levrie and Vaishnavi Sundararajan for their substantial and very much appreciated efforts to proofread the final version of the manuscript. Peter van Emde Boas has indefatigably provided details on the history of the Consecutive Numbers riddle, and has much encouraged us in writing this book. We wish to thank Allen Mann, Springer, for his encouragement and for getting us started on this project. Nicolas Meyer from the ENS des Mines in Nancy found an embarrassing error in a light bulb protocol when Hans gave a course there, only a few weeks before we handed over the manuscript. He is one of many. If one were to go back all the 25 years of teaching logic and puzzles at colleges, universities, and summer schools, a much longer list of thanks to students and colleagues would be appropriate: by making an example of one, we wish to thank them all. No doubt, there will still be many remaining errors. They are all the responsibility of the authors.

Nancy, France, and Groningen,   Hans van Ditmarsch
the Netherlands   and Barteld Kooi
25 December 2014

Contents

1

Consecutive Numbers

Anne and Bill get to hear the following: "Given are two natural numbers. They are consecutive numbers. I am going to whisper one of these numbers to Anne and the other number to Bill." This happens. Anne and Bill now have the following conversation.

* *Anne: "I don't know your number."*
* *Bill: "I don't know your number."*
* *Anne: "I know your number."*
* *Bill: "I know your number."*

First they don't know the numbers, and then they do. How is that possible? What surely is one of the two numbers?

The natural numbers are the numbers 0, 1, 2, 3, etc. Numbers are consecutive if they are one apart. It is important for the formulation of the riddle that Anne and Bill are simultaneously aware of this scenario, and also know that they both are aware of this scenario, etc. Therefore, they are being spoken to, instead of, for example, both receiving written instructions. It is therefore too that the numbers are whispered into their ears—the whispering creates common knowledge that they have received that information. We can imagine the setting of this riddle as Anne, Bill, and the speaker sitting round a table, such that the speaker has to lean forward to Anne in order to whisper to her, and subsequently has to lean forward to Bill and whisper to him.

1.1 Which Numbers Are Possible?

We solve the riddle by analyzing the developing scenario piecemeal. The first bit of information is as follows:

* Given are two natural numbers.

We do not know yet what these numbers are, but apparently there are two relevant variables: the number x that Anne is going to hear and the number y that Bill is going to hear. The question is then to determine the pair (x, y). We also know that x and y are *natural numbers*: 0, 1, 2, etc. So, the possible pairs are $(0, 0)$, $(0, 1)$, $(1000, 243)$, etc. Of course there are infinitely many such pairs. The state space consisting of all such pairs looks as follows—to simplify the representation we write xy instead of (x, y), and for convenience we order the number pairs in a grid.

⋮	⋮	24	34	44
03	13	23	33	43
02	12	22	32	42
01	11	21	31	⋯
00	10	20	30	⋯

The number pair $(1, 2)$ is different from the number pair $(2, 1)$: The first of each pair is the number that Anne is going to hear, whereas the second of each pair is the number that Bill is going to hear. In $(1, 2)$, Anne is going to hear 1, and in $(2, 1)$ she is going to hear 2.

The next bit of information is that

* They are consecutive numbers.

This means that the only possible number pairs (x, y) are those where $x = y + 1$ or $y = x + 1$. Hence, only these pairs remain:

1.2   What Anne and Bill Know

So far, your perspective, as reader, is the same as Anne's and Bill's: The numbers are natural numbers, and they are consecutive. These are all the possibilities that we have to take into account. We cannot distinguish among these pairs. The next bit of information makes Anne's and Bill's perspective different from your perspective as reader:

* "I am going to whisper one of these numbers to Anne and the other number to Bill." This happens.

Suppose that the whispered numbers were 5 to Anne and 4 to Bill. After Anne hears 5, she knows that Bill's number is 4 or 6. She can rule out all number pairs except $(5, 4)$ and $(5, 6)$. Bill's view of the situation is different from Anne's. He hears 4. After that, the remaining number pairs from his perspective are $(5, 4)$ and $(3, 4)$. You, the reader, cannot rule out any number pair! But you still have learnt something, namely what Anne and Bill learnt about any number pair and about each other. We can make the information change visible in the given set of consecutive number pairs: We can indicate which pairs are indistinguishable for Anne or for Bill after the whispering has taken place. A visual means is to link such pairs by an edge labeled with *a* for Anne, or *b* for Bill. We get:

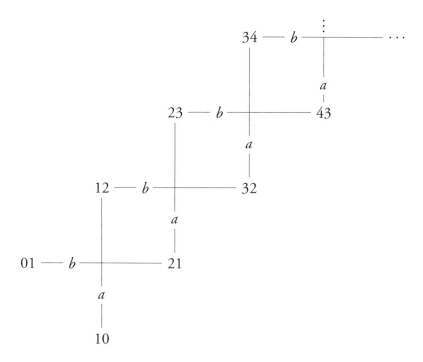

We might as well have the figure topple over a bit to save space on the page:

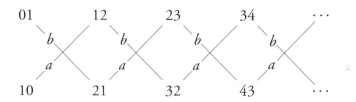

In fact, we simply have two infinitely long chains of number pairs, with alternating labels. So, alternatively, just one of those is as follows:

$$10 \longrightarrow a \longrightarrow 12 \longrightarrow b \longrightarrow 32 \longrightarrow a \longrightarrow 34 \longrightarrow b \longrightarrow \cdots$$

Anne's and Bill's perspectives are now different from each other and also from your perspective as a reader. Before the whispering action, all number pairs were equally possible for Anne, for Bill, and for you. After the whispering, all number pairs remain possible for you—they can equally well be 3 and 4, or 5 and 4, or 89 and 88—but for Anne and Bill this is no longer the case: If Anne were to have 3, she would know that the other number cannot be 88, but only 2 or 4. What you have learnt as a reader is that Anne and Bill now have this knowledge.

1.3   Informative Announcements

A figure such as the above we call a *model* of the description of the initial state of the riddle. We changed the model piecemeal with every new bit of information in the problem description. There were two sorts of changes: eliminating number pairs (for example, those number pairs that were not consecutive numbers), and indicating which number pairs could be distinguished by Anne and by Bill (for example, that Anne can distinguish $(2, 3)$ from $(5, 6)$, but not $(2, 3)$ from $(2, 1)$). Next on our list of problem-solving activities is to convert each announcement by Anne and Bill into some such model transforming operation. In this riddle, all further changes are of the first kind: elimination of number pairs. The crucial aspect here is that we do not treat Anne's announcement differently from the "announcements" of the anonymous speaker who informs Anne and Bill in the beginning. Anne and Bill both hear their own announcements, and know from one another that they both hear what they say, and so on. And also, you as a reader can be said to be "hearing" the announcements: You have to imagine yourself as silent bystander present at the interaction between the initial speaker and Anne and Bill, and at their subsequent announcements. Let us take the first announcement:

* Anne: "I don't know your number."

When would Anne have known what Bill's number is? Suppose Anne had heard 0. She knows that Bill's number is one more or one less than her own. It cannot be -1, as this is not a natural number. Therefore, the only remaining possibility is that Bill's number is 1. So, Anne then *knows* that Bill has 1. However, as she says, "I don't know your number," we can rule out the number pair $(0, 1)$. And not just we, but also Bill. The change is public (for Anne and for Bill), because Anne said it aloud. If she had, for example, written it on a piece of paper, this might have created uncertainty in her whether the message had reached Bill, or uncertainty in Bill whether Anne knew that the message had reached him, and so on. The message would not have been public. Given that the change is public, the result is as follows:

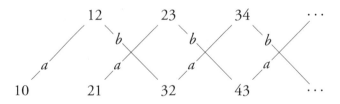

It is now crucial to observe that this is a different model, and that it may therefore satisfy different propositions. Propositions that were false before may now be true, and propositions that were true before may now be false. This will explain why saying, "I don't know your number" now and "I know your number" later only appears to be in contradiction, but is not really a contradiction. These observations are about different information states of the system. The announcements help us to resolve our uncertainty about what the number pair is. Similarly, it will help Anne and Bill to resolve their uncertainty. We continue our analysis by processing the next announcement:

* Bill: "I don't know your number."

When would Bill have known what was Anne's number? There are two possibilities. In the first place, Bill would have known Anne's number if the number pair had been $(2, 1)$. If Bill has 1, then he can imagine Anne to have 0 and 2. Given that 0 is no longer possible after Anne's (first) announcement, only 2 remains. So, Bill then knows that Anne's number is 2. But there is yet another pair where Bill would have known Anne's number, namely $(1, 0)$. Now, just like Anne in the case of $(0, 1)$, Bill would have known that Anne has 1 because -1 is not allowed. Because Bill said, "I don't know your number," neither of these two pairs can be the actual pair. The resulting situation is as follows:

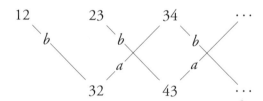

This brings us to the third announcement:

* Anne: "I know your number."

We can see in the model that this is true for the number pairs $(2, 3)$ and $(1, 2)$, as there is then no alternative left for Anne. We can alternatively see this as the conclusion of a valid argument. For example, for the pair $(2, 3)$:

> If Anne has 2, then she now knows that Bill has 3, because, if Bill were to have 1, he would have said in the second announcement that he knew Anne's number. But he did not.

All other number pairs have become impossible because of her announcement. The resulting model is therefore,

12   23

This depicts that if the numbers are 1 and 2, then Anne and Bill know this, know from one another that they know this, etc. It is common knowledge between them. If the numbers are 3 and 2, then they also have common knowledge of the numbers. Although both $(1, 2)$ and $(2, 3)$ are in the model, this does not mean that if the numbers are 1 and 2, then Anne and Bill also consider it possible that they are 2 and 3: There is no link for a or for b in the model. But you, as a reader, cannot determine which of the two pairs must be actually the case. We now get to the last announcement:

* Bill: "I know your number."

This proposition is already true for both remaining number pairs. Therefore, nothing changes. We could also have said: This last announcement was not informative. Anne already knew that Bill knew her number, and they both knew this.

This solves the riddle. All four announcements were truthful. The contradiction between "I don't know your number" and "I know your number" is not a contradiction in the riddle, because these announcements are made at different moments. What was true before can be false later. After the four announcements, the remaining number pairs are $(1, 2)$ and $(2, 3)$. You cannot choose between these two pairs. But the number 2 occurs in both pairs, and is therefore certainly one of the two numbers.

1.4   Versions

Puzzle 1 *Suppose that the actual numbers are neither 1 and 2, nor 2 and 3, but 4 and 5. The four announcements can no longer all be made truthfully. What is going wrong? How often does "I don't know your number" have to be repeated for Anne and Bill to get to know the other number, and by whom?*

Puzzle 2 *An alternative presentation of the riddle is as follows:*

> *Anne and Bill* **each have a natural number on their forehead.** *They are consecutive numbers. Anne and Bill now have the following conversation.*
>
> * *Anne: "I don't know* **my** *number."*
> * *Bill: "I don't know* **my** *number."*
> * *Anne: "I know* **my** *number."*
> * *Bill: "I know* **my** *number."*
>
> *What difference does this formulation make for the solution?*

Puzzle 3 *Suppose that the numbers are not consecutive, but* **two** *apart. So, the riddle will be as follows:*

> *Anne and Bill get to hear the following: "Given are two natural numbers. The numbers are two apart. I am going to whisper one of these numbers to Anne and the other number to Bill." This happens. Anne and Bill now have the following conversation.*

> * *Anne: "I don't know your number."*
> * *Bill: "I don't know your number."*
> * *Anne: "I know your number."*
> * *Bill: "I know your number."*

What does the model look like in this case, and how it is transformed due to the announcements? And what if the numbers are *m* apart, where *m* is a natural number?

Puzzle 4 *Suppose there is a third person playing the game, Catherine. Now, the riddle is:*

> *Anne, Bill,* **and Catherine** *each have a natural number on their forehead. They are consecutive numbers. Suppose, for example, that the numbers are 3, 4, and 5 (respectively). What sort of conversation is possible between Anne, Bill, and Catherine, on knowledge and ignorance of each other's number, in order to find out their own number?*

Puzzle 5 *Anne and Bill have a natural number on their forehead. It is known that the sum of these two numbers is equal to 3 or 5. Anne and Bill may now consecutively announce whether they know their own number. Show that they can have the following conversation:*

* *Anne: "I don't know my number."*
* *Bill: "I don't know my number."*
* *Anne: "I know my number."*
* *Bill: "I know my number."*

(After Conway et al. (1977); see the history section below.)

1.5 History

An original source for the riddle is found straight at the beginning of *A Mathematician's Miscellany* by Littlewood (1953, p. 4):

> There is an indefinite supply of cards marked 1 and 2 on opposite sides, and of cards marked 2 and 3, 3 and 4, and so on. A card is drawn at random by

a referee and held between the players A, B so that each sees one side only. Either player may veto the round, but if it is played the player seeing the higher number wins. The point now is that every round is vetoed. If A sees a 1 the other side is 2 and he must veto. If he sees a 2 the other side is 1 or 3; if 1 then B must veto; if he does not then A must. And so on by induction.

In the Littlewood version, there is no "solution" (every round is vetoed), and the synchronization is left open to interpretation (who vetoes first?). But a player seeing number x on one side of the playing card is uncertain if the number on the other side is $x + 1$ or $x - 1$. Only when a player is seeing the number 1 can he be certain about the other number, namely that it is 2 (the number 0 is ruled out). This version is also treated, slightly differently, by Gardner (1977):

> You are one of two contestants in the following game: An umpire chooses two consecutive positive integers entirely at random and writes the two numbers on slips of paper, which he then hands out randomly to the two players. Each looks at their number and either agrees or disagrees to play. If both players agree, the person with the higher number must pay that many dollars to their opponent. You only agree to play when the expected payout favors you. Obviously, you would agree if your number was 1. For what other values should you agree to play?
>
> Assume infinite resources for payouts. I.e. it does not matter how high the numbers are, the payment can be made.

A far more general version of the riddle is found in *A Headache-Causing Problem* by Conway et al. (1977). This is a contribution to an honorary volume "presented to Hendrik W. Lenstra on the occasion of his doctoral examination." The treatment is light, for example, the initials of the third author are "U.S.S.R." This is because Paterson and Conway discussed the riddle while waiting in transit on Moscow airport (as van Emde Boas recently found out).

> There are n persons, all having a natural number on their forehead. It is known that the sum of these n numbers is equal to one of at most n possible given numbers. The n players may now consecutively announce if they know their own number, until one of them says that he or she knows it. Prove that this will happen eventually.

The last publication in this series of original sources is then *The Conway Paradox: Its Solution in an Epistemic Framework* by van Emde Boas, Groenendijk, and Stokhof, orginally presented at the Amsterdam Colloquium in 1980, afterwards published in *Mathematical Centre Tract No. 135* in 1981, and finally published in book format in (van Emde Boas et al. 1984). This publication is an important precursor of dynamic epistemic logic. It also provides a very accurate historical section, on which this overview is based. After their publication, the consecutive numbers riddle became known as the *Conway paradox*. Yet another nice story comes with that: It is curious to observe that the consecutive numbers riddle, even though it is now known as the Conway paradox, is *not* a special case of the problem described in Conway et al. (1977), so that "Conway paradox" is actually a misnomer for the consecutive numbers riddle, as van Emde Boas confirms.

For example, if Anne has 3 on her forehead and Bill 2, that indeed involves uncertainty by two players about two numbers, and therefore also about two sums of numbers, but, unlike the Conway version, this is uncertainty about more than two sums: Anne is uncertain if the sum is 5 or 3, whereas Bill is uncertain if the sum is 5 or 7. And, of course, Bill is uncertain whether Anne is uncertain between sums 5 and 3, or between sums 7 and 9, and so on. An infinity of sums plays a role.

On a more abstract level (no doubt in the mind of van Emde Boas et al. at the time), there is of course a correspondence. See Puzzle 5.

2

Hangman

At a trial a prisoner is sentenced to death by the judge. The verdict reads "You will be executed next week, but the day on which you will be executed will be a surprise to you." The prisoner reasons as follows. "I cannot be executed on Friday, because in that case I would not be surprised. But given that Friday is eliminated, then I cannot be executed on Thursday either, because that would then no longer be a surprise. And so on. Therefore the execution will not take place." And so, his execution, that happened to be on Wednesday, came as a surprise.

So, after all, the judge was right. What error does the prisoner make in his reasoning?

The prisoner's argument is very convincing. At first sight it seems as if it cannot be refuted at all. Still, the conclusion cannot be right. The prisoner rules out that the hanging will be on Thursday, and that it will be on Wednesday, and so on, but in fact the hanging is on Wednesday. Is it not easy to make clear where the error is. And therefore it is indeed called a paradox. To find the error in the prisoner's reasoning, we first have to define what a "secret" is. Because initially the day of the hanging is a *secret*.

2.1 How to Guard a Secret?

The best way to guard a secret (like who you are in love with) is not ever to tell it to anyone. That is easier said than done. If your head is filled with the secret, it can happen to fall out of your mouth before you know it. And then it is no longer a secret. Someone might ask you why you are staring out of the window all the time, focusing on the horizon. You can then of course say that this is because you are guarding a secret. But that makes it less secret. If you really want to guard a secret, you had better not ever talk about it, because if you do, then you risk that the secret will be discovered.

It is also a bad idea to talk to yourself about your own secret. A classic of that kind (renewedly popular from the TV series *Once Upon a Time*) is the character Rumpelstiltskin in the Grimm Brothers fairy tale by the same name (1814). The queen promised her first-born child to Rumpelstiltskin. There is an escape clause: If the queen guesses correctly Rumpelstiltskin's name, then she can keep her child. She can guess three times. The first two guesses are incorrect. The tension rises. The queen's messenger now tells her that he saw in the forest, from a hidden place behind some bushes, a funny guy who was dancing while singing, loudly:

Heute back ich, morgen brau ich,
übermorgen hol ich der Königin ihr Kind;
ach, wie gut dass niemand weiß,
dass ich Rumpelstilzchen heiß!

Fortunately, the messenger understood German and could translate this into

Today I'll bake; tomorrow I'll brew,
Then I'll fetch the queen's new child,
It is good that no one knows,
Rumpelstiltskin is my name.

It was indeed Rumpelstiltskin who was singing this song, and so the queen finds out his name, and the third time her guess is correct: "Your name is Rumpelstiltskin." If only he had kept his mouth shut, it would have remained a secret.

The funny thing is, that the last two sentences, in a more convenient phrasing "Nobody knows that my name is Rumpelstiltskin," become false because Rumpelstiltskin is singing it. After this, it is no longer the case that nobody knows that his name is Rumpelstiltskin. The messenger now knows. This phenomenon is quite special. Apparently, it is possible to say something ("Nobody knows that I am in love with Stephanie") but because I am saying it, it becomes false. (In no time everyone, including Stephanie, knows that I am in love with her.) Usually when we say something, it remains true after we say it. But in exceptional cases, this is apparently false.

What is the relationship between hangings and fairy tales? The day of the hanging is a secret guarded by the judge, and the prisoner can only guess what the exact day is. The judge does not tell which day it is. What does it mean that the judge says that the day of the hanging will be a surprise? A surprise is something unexpected, it is something happening that you did not see coming. In the reasoning of the prisoner, "surprise" is entirely interpreted in terms of *knowledge*. The hanging is a surprise, because the prisoner does not *know* the day of the hanging in advance. A secret is no secret anymore if you are telling it to someone, just as for the secret of Rumpelstiltskin. Similarly, a surprise is not a surprise anymore if you announce it. If you want to surprise someone with a big bunch of roses, then you should not let it appear from your behavior. If you say, "I am going to surprise Stephanie tomorrow with a big bunch of roses," then the surprise is lost when she hears about it. If Rumpelstiltskin says, "Nobody knows that my name is Rumpelstiltskin," then someone may get to know it.

2.2   A Bridge Too Far

When the judge says that the day of the hanging will be a surprise, he risks spoiling the surprise. If he had not said anything about the hanging, not even that it was going to be next week, it would not have mattered; surely the prisoner would then have been surprised by the hanging.

The error that the prisoner seems to make in his reasoning is that he does not realize that the judge may have spoilt the surprise by announcing it. Before the judge is saying that the day will be a surprise, the prisoner considers it possible that the hanging will take place on one of Monday, Tuesday, Wednesday, Thursday, or Friday. Now suppose nothing else has been announced about the day of the hanging. The prisoner would then know on Thursday night that the hanging will be on Friday. The hanging would then not be a surprise. On all other days, it would be a surprise. This, the judge also knows. But by saying that to the prisoner, he spoils the surprise. His announcement rules out that the hanging will be on Friday. Therefore, if the prisoner had not yet been hanged by Wednesday night, he could by that time have concluded that the hanging must be on Thursday. So now Thursday is special, instead of Friday.

However, the prisoner takes the argument further—and too far: He assumes that even after the judge's announcement, the day of the hanging remains a surprise. And, therefore, he thinks he can rule out not only Friday but also Thursday, and Wednesday, and Tuesday, and Monday. But that is carrying it too far. Only Friday can be ruled out.

In fact, the hanging is on Wednesday. So, if the prisoner would not get more information, that would still be a surprise.

Let us illustrate this by constructing models. We assume that initially the prisoner only knows that there will be a hanging some day next week (a working day: Monday to Friday). So, this is before the judge announces that the day of the hanging will be a surprise. In that case, how will the prisoner's information change with the passing of that coming week? Below we can see this depicted for the different days that the hanging can take place. Two events may reduce the uncertainty for the prisoner: Nightfall will rule out that the hanging is on the current day and thus reduces the uncertainty, but the hanging itself will confirm that it is on the current day and thus also reduces the uncertainty.

Hang on, what does it mean for a prisoner who has been hanged and who is dead, to know on Friday that he has been hanged on Thursday? Dead prisoners do not know anything. True enough, but this is an artifact of our setting of the riddle! In another version, the riddle concerns a surprise *exam* given by a schoolmaster to his pupils. Then, on Friday you will still know that the exam has been on Thursday. We can also imagine ourselves, as problem solvers, to be the agents observing the scenario and whose knowledge is being modeled. The problem solver will still know on Friday that the prisoner has been hanged on Thursday.

What is remarkable in these different scenarios is that there is only one occasion where the actual hanging does remove the uncertainty about the day of the hanging, for the prisoner. Namely, when the hanging is on Friday. Because (only) on that occasion the prisoner can determine the night before the hanging that the hanging will take place on Friday. So, for the prisoner there is only one day where the hanging will not be a surprise: Friday. If the judge announces that the hanging will be a surprise, this then rules out that the hanging is on Friday.

After the judge's announcement it is not necessarily so that the hanging will be a surprise. But there is now another scenario in the picture above where the hanging will not be a surprise, namely where Friday has been eliminated and where the hanging will be on Thursday. The prisoner does not know this in advance but knows that the hanging will be on Thursday when Wednesday night falls.

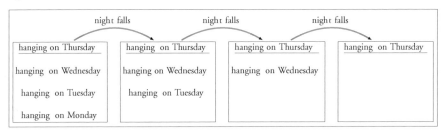

2.3   Versions

Puzzle 6 *Suppose that the judge has answered the question "On which day?" by "That will* not *be a surprise." On which day will the hanging then take place?*

A different wording of the riddle is not about a judge surprising a prisoner with the day of a hanging, but about a schoolmaster or teacher surprising a class with the day of an examination: "You will get an exam next week, but the day of the exam will be a surprise for you." Then, of course, the class only learns that the examination will not be on Friday. It is therefore also known as the surprise exam paradox. We now discuss a further version of that.

Puzzle 7 *Suppose the teacher, Alice, had only said that the exam would take place next week, but without saying that the exam would come as a surprise.*
During lunch break, her pupil Rineke walks past the staffroom and overhears the teacher saying to a colleague, "I am going to give my class an exam next week, and the day of the exam will be a surprise to them." The teacher did not realize that Rineke was overhearing her. What can Rineke conclude on the basis of this information about the day of the examination?

But the plot thickens. Because after lunch break Rineke says to the teacher, "I heard you say that the day of our exam next week will come as a surprise." The teacher confirms this. However, later that day, while getting her parked bike from the bikeshed, the teacher meets the staffroom colleague again, who is about to go home as well, and tells him that Rineke had overheard them earlier that day, and says, "But the day of the exam will still come as a surprise!" Unfortunately, Rineke overhears this again. What can Rineke now conclude about the day of the exam?

2.4 History

During the Second World War, the Swedish mathematician Lennart Ekbom overheard a radio message announcing a military training exercise next week. The training exercise would, of course, come as a surprise. It occurred to him that this message seemed paradoxical (Kvanvig 1998; Sorensen 1988, p. 253). The paradox was then published by O'Connor (1948). One of the responses to this publication then mentions that the exercise could still take place (Scriven 1951), what makes it even more paradoxical.

There are many versions of the paradox, the best known is the "surprise exam" version where a schoolmaster announces to his class that an examination will be given next week, but that the day will be a surprise (this first appeared in Weiss 1952). The "hangman" version of our presentation first appeared in Quine (1953).

The treatment of the puzzle differs depending on how "surprise" is interpreted. This can be done in many different ways. It can be in terms of derivability (the precise day *does not follow from* what the judge says). This approach was followed by Shaw (1958). But of course "surprise" can also be interpreted as "ignorance": lack of knowledge. This is what we have done here. "The prisoner will be surprised" then means that the prisoner does not *know* in advance when the hanging will take place.

Since 1948, more than 100 publications have appeared on the hangman paradox. They contain even more interpretations. A detailed overview of treatments of the paradox and its history is given by Sorensen (1988).

It is remarkable that all this "scientific work" has not resulted in a universally accepted solution of the paradox. Chow (1998) even calls it a meta-paradox:

> The meta-paradox consists of two seemingly incompatible facts. The first is that the surprise exam paradox seems easy to resolve. [...] The second (astonishing) fact is that to date nearly a hundred papers on the paradox have been published, and still no consensus on its correct resolution has been reached.

The solution given in this chapter is based on the work of Gerbrandy (1999, 2007). It is also treated by van Ditmarsch and Kooi (2005, 2006).

3
Muddy Children

A group of children has been playing outside and they are called back into the house by their father. The children gather round him. As one may imagine, some of them have become dirty from the play. In particular: they may have mud on their face. Children can only see whether other children are muddy, and not if there is any mud on their own face. All this is commonly known, and the children are, obviously, perfect logicians. Father now says: "At least one of you is muddy." And then: "Will those who know whether they are muddy step forward." If nobody steps forward, father keeps repeating the request. At some stage all muddy children will step forward. When will this happen if m out of k children in total are muddy, and why?

This is a puzzling scenario. If there is more than one muddy child, all children see at least one muddy child, so they know that there is at least one muddy child. Father then says something that everyone already knows. If that is so, why say it? And why, after making the request to step forward, would he repeat this request? If nobody responds by stepping forward, what difference would it make to repeat the request? To understand that this makes a difference, we look at a simpler puzzle first.

3.1 Muddy or Not Muddy, That is the Question

Puzzle 8 *Alice and Bob are coming home from playing outside. Their father notices that they have been playing with mud, because Bob has mud on his face. They can only see mud on each other's face, but not on their own face. Of course, you can find out by looking in a mirror. Father now says "One of you has mud on his face." Bob now leaves and washes his face. However, he did not look in a mirror. How did he find out that he is muddy?*

To solve such puzzles we have to assume that all children are geniuses (what every parent will happily confirm): They are perfect logicians. Also, we assume that the father and his children are always speaking the truth, and that they have complete confidence in each other speaking the truth. If a child has mud on the face, we might as well say that *the child is muddy*.

Father told Alice and Bob that one of them is muddy. Bob can see that his sister has no mud on her face. Therefore, he must have mud on his face. He did not have to look in the mirror for that.

Puzzle 9 *The next day Alice and Bob are playing outside, again, but now they both are muddy. When coming home, father says, again, "At least one of you is muddy." Father now asks Bob, "Do you know whether you are muddy?" Bob responds, "No, I don't know." Then he asks Alice, "Do you know whether you are muddy?" And Alice responds, "Yes, I know. I am muddy." How is it possible that Alice knows but Bob not?*

Let us recall that "you know whether you are muddy" means "you know that you are muddy or you know that you are not muddy." In many languages there is this kind of difference between "knowing that" and "knowing whether!" (Savoir si, savoir que; weten of, weten dat; saber si, saber que; . . .)

The situation of Alice and Bob seems completely symmetrical. They are both muddy (and, really, both only on their forehead which they obviously cannot see, there are no tricks), they both get the same information from their father, and they have even both been asked the same question. That makes it even more puzzling that Bob's answer is different from Alice's.

The only difference between Bob and Alice is that Bob was asked first, whereas Alice was asked second. Therefore, Alice heard Bob say that he does not know whether he is muddy. This piece of information is crucial to understand the different responses of the two children. Before Bob answers, Alice considers it possible that she is not muddy. For Alice, it could have been the same situation as before (in Puzzle 8), wherein only Bob was muddy. In that case, Bob would have seen that Alice is not muddy, and he would then have concluded that he has mud on his forehead. As he says that he does not know, he therefore did not draw that conclusion, and that can only be because he sees that Alice is muddy. Alice can come to this conclusion by her own reasoning, and therefore concludes that she must be muddy. So, she can answer the question positively.

After Alice's response, Bob remains uncertain whether he is muddy, because Alice would also have said that she knows that she is muddy if she had seen that Bob was not muddy. The situation would then have been as in the previous Puzzle 8, but with only Alice being muddy instead of only Bob being muddy.

Puzzle 10 *The day after that, Alice and Bob have been playing again, at least one of them has become muddy, and father says again, "At least one of you is muddy." He now asks Alice, pointing to Bob, "If I were to ask Bob if he knows whether he is muddy, what would be his answer?" Alice answers, "He will answer, 'I don't know'." Who is muddy?*

Alice is muddy and Bob is not muddy! Suppose that Bob was muddy. Then Alice would see that Bob is muddy. Now she cannot know in that case whether she is muddy herself. If she were not muddy, then Bob would know that he is muddy. But if she were muddy, then Bob would not know whether he is muddy. Therefore, Alice would not know if Bob knows whether he is muddy in that case, and therefore she would not have been able to say that Bob does not know whether he is muddy. But she said, "He will answer 'I don't know'." In other words, Alice knows that Bob does not know whether he is muddy. Therefore, the supposition that Bob was muddy must be wrong. And therefore, Bob is not muddy. As there was at least one muddy child, Alice is muddy and Bob is not muddy.

Puzzle 11 *Consider another family, consisting of two children: Anne and Bill. Bill is blind. So, he cannot see his own face, but he can also not see Anne's face. They play outside and both get mud on their face. After they have come home, father says, "At least one of you is muddy." Anne asks her father, "Am I muddy?" Even before father answers the question, Bill leaves to clean his face. Why?*

Bill is blind, but he can imagine that he is his sister, who can see. In order to solve this puzzle, we have to assume that Anne only asks questions to which she does not know the answer. So, when she is asking whether she is muddy, this gives away that she does not know whether she is muddy. Her question is like an announcement that she does not know whether she is muddy. If Bill was clean, then Anne would know that she is muddy, because at least one child is muddy. But she does not know. Therefore, Bill must be muddy. Because Bill can reason about what Anne knows and does not know, he can also draw this conclusion and therefore will go clean his face.

3.2   Simultaneous Actions

Up to now, only a single child responded, or the children could respond in turn. But in the original version of the riddle, they are asked to do something simultaneously: To step forward, or not. There is a problem with this stepping forward: when nothing happens after the father makes his request, is that because some or all children are still thinking about whether they should step forward, or is that because they have decided not to step forward? We need to synchronize the action of stepping forward or not, to eliminate this ambiguity. This is the role of father clapping his hands, in the following puzzles: There is now a precise moment, namely exactly *then,* when every child has supposedly finished thinking about what to do. Nobody stepping forward at that moment really means that *nobody* knows whether he or she is muddy, and some children stepping forward at that moment means that *only* those children know whether they are muddy.

Puzzle 12 *As before, Alice and Bob played outside. They both got mud on their face. And also as before father informs them that at least one of them is muddy. Then he says, "I will clap my hands. If you know whether you are muddy, please step forward." He claps his hands, but neither Alice nor Bob steps forward. He then repeats what he said before, "I will clap my hands. If you know whether you are muddy, please step forward." This time, when he claps his hands, Alice and Bob both step forward.*

Explain why this is possible.

This puzzle is already looking somewhat like the puzzle at the start of this chapter. Father repeats his request after neither of the children responds the first time. Why does he repeat it? By not moving when father claps his hands, the children indicate that they do not know whether they are muddy. Because Alice does not step forward, Bob learns that she does not know whether she is muddy. Simultaneously, because Bob does not step forward, Alice learns that he does not know whether he is muddy. So, by both not making a move, they are in fact communicating with each other. They are not really doing nothing. Alice and Bob both see one muddy child. Before they both did not step forward, they both considered it possible that they themselves were not muddy. But afterwards, neither considers this possible anymore. Alice and Bob can both draw this conclusion about the reasoning of the other, and therefore they now both know that they are muddy.

We do this again for three muddy children. It remains possible to explain all this in words. But we will now also use pictures to explain how the children are reasoning.

Puzzle 13 *Alice, Bob, and Caroline are coming home from playing outside and all three have mud on their face. Father tells them that at least one of them is muddy and then says, "In a moment I will clap my hands. If you know whether you are muddy, please step forward." He claps his hands. Nothing happens. He repeats this twice. The third time he claps his hands, all three children step forward. Explain how this is possible.*

To solve this puzzle, we depict all possible situations. We determine a situation by stating for each child if it is muddy or not muddy. ("Not muddy" is the same as clean.) There are therefore eight situations. In the figure below, a situation is represented by three bits. A bit has the value 0 or 1. The first bit has value 1 if Alice is muddy, and it has value 0 if Alice is not muddy. The second bit stands for the mud on Bob's face and the third bit for the mud on Caroline's face. For example, the situation wherein Alice and Bob are muddy but Caroline is not muddy is represented by 110.

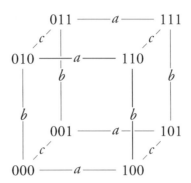

Some of the possible situations have been linked by lines with labels. If two situations are linked with an *a*-'el, then Alice (*a*) cannot distinguish these situations. For example, Alice cannot distinguish 011 from 111: She is uncertain if only Bob and Caroline are muddy or if they are all three muddy. Therefore, these two situations are *a* linked. Similarly, we link other states that are indistinguishable for Anne, and states that are indistinguishable for Bill (*b*) or for Caroline (*c*).

It is tempting to think that not all possible situations matter to solve the problem. For example, if all three children are muddy, then no child considers it possible that none of them are muddy. (There is no line connecting 111 and 000.) But this is wrong. Situation 000 still matters. Let us assume that the situation is indeed 111. Then Alice cannot rule out that she is clean. But if that were so (situation 011), then Bob cannot rule out that he is clean (situation 001). In other words, Alice cannot rule out that Bob cannot rule out that Alice and Bob are both clean. Now if that were the situation (001), then Caroline cannot rule out that they are all three clean (000). Therefore, Alice cannot rule out that Bob cannot rule out that Caroline cannot rule out that all three children are clean. Such a phrase is hard to grasp intuitively. But in the picture above, it is easy to grasp visually: There is an *a* line from 111 to 011, there is a *b* line from 011 to 001, and there is a *c* line from 001 to 000. So these states are connected by a finite chain of links (possibly) labeled with different agents.

Now what happens if father says that at least one of them is muddy? After that announcement, 000 is no longer considered possible by *any* of the three agents Alice, Bob, and Caroline, and they all know that they no longer consider that situation possible, and so on. The situation 000 is no longer considered by anyone, by any chain of reasoning. We can therefore remove it from the picture. We get:

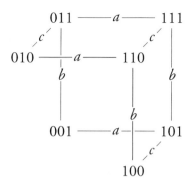

The situations 001, 010, and 100 in this picture are kind of special. In those cases, there is one child for whom the situation is different from all other situations (there is no line labeled with that child and connecting it to another situation). In 001, this is Caroline; in 010, this is Bob; and in 100, this is Alice. In the picture for the previous state of information, they could not rule out 000, but now they can. This means that if the situation is 001, then Caroline knows that she is muddy; if the situation is 010, then Bob knows that he is muddy; and if the situation is 100, then Alice knows that she is muddy.

Father now claps his hands and nothing happens. This apparent lack of action is informative. It means that no child knows whether it is muddy. It rules out the situations 100, 010, and 001, because if one of those had been the case, then Alice, Bob, or Caroline, respectively, would have stepped forward. We can therefore further revise the figure by deleting these possible situations. We now know that these are not the actual situation.

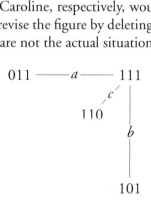

Now consider the situations 011, 101, and 110. In these situations, there are no outgoing lines for two of the three children. In other words, for those children that is now the unique remaining possibility. For example, from 011, there are no b and c lines connecting it to other situations. This means that Bob and Caroline now know that they have mud on their face. We can retrace in English, why this must be so. In 011, Bob sees that Caroline is muddy and that Alice is clean. Initially, it remained possible that only Caroline was muddy.

But if that had been the case, Caroline would have stepped forward. She did not. Therefore, Bob concludes that Caroline sees a muddy child. That must be himself. Caroline can similarly draw the conclusion that she is muddy.

Again father claps his hands, and again nothing happens. Again no child knows whether it is muddy. The situations 011, 101, and 110 that we just discussed, where the two muddy ones would now have stepped forward, can therefore be ruled out. Only one situation remains, 111. The picture becomes very simple:

111

There are no lines at all, there are no alternative situations to consider, and therefore all three children know that they are muddy. When father now claps his hands, for the third time, all three muddy children will step forward.

This does not merely solve Puzzle 13, but a whole range of similar puzzles involving three children. If not three but two had been muddy, then those would have stepped forward after the second time father claps his hands. And if only one had been muddy, clapping once would have been sufficient. Let us do one more general case.

Puzzle 14 *Imagine a large family consisting of 26 children: Alice, Bob, . . . , Yolanda, and Zacharias. They have all become muddy while playing outside. Father says that at least one of them is muddy, and then says, "In a moment I will clap my hands. If you know whether you are muddy, please step forward." He then repeats this request, and including the first time that he made it, the request is finally made 26 times. At that moment, all 26 muddy children will step forward. Explain why this is possible.*

And if there had only been 20 muddy children? What would then have happened?

For the solution of this problem, we cannot use a picture any more. It would be too large. For 26 children, there are initially 2^{26} possible situations, i.e., 67,108,864. We have to think of something else to solve the problem. Let us do this systematically.

If there had been only one muddy child, that child would not see any other muddy children. So, that child would know that it is muddy when it has obtained the information that there is at least one muddy child. With father's first clap of hands, that child will step forward.

If there are two muddy children, then each muddy child will see one other muddy child. If no child will step forward after the first clap of hands, then both muddy children can conclude that, apart from the single muddy child they see, there must be yet another muddy child. They must therefore themselves be that muddy child. So, the two muddy children will both step forward the second time the father claps his hands.

The same sort of argument holds for three children. Then, all three muddy children step forward at the third clap of hands. And so on. If there are 20 muddy children, they will step forward at the 20th clap, and if there are 25 muddy children, then at the 25th time. Now as there are 26 children that are all muddy, any child sees 25 muddy children and thinks: If I am clean, then all the others will step forward after the 25th clap of hands. But if they do not, then I must be muddy too and there are 26 muddy children. All 26 children come to that conclusion at the same time, and therefore they all step forward after the 26th clap of hands.

We can now solve the puzzle at the start of the chapter. If there are k children, of which m are muddy, then the m muddy children will step forward after the mth request to do so. This requires a proof by natural induction.

3.3 Versions

Puzzle 15 *This is a version of Puzzle 13. We recall that all three children are muddy. After father tells them that at least one of them is muddy, they all say, at the same time, "I already knew." What do the children learn from this announcement?*

Suppose father now continues by saying, "In a moment I will clap my hands. If you know whether you are muddy, please step forward." How often does father have to make this request before all children step forward?

Puzzle 16 (Washing muddy children) *Alice, Bob, and Caroline are coming home after playing outside. Their father informs them that at least one of them is muddy. In plain view of all, he now wipes Alice's face with a clean towel (he makes Alice clean), and then says, as before, "In a moment I will clap my hands. If you know whether you are muddy, please step forward." Father then claps his hands, and again, and again.*

1. *What will happen if only Alice was (initially) muddy? What do Bob and Caroline learn from this scenario?*
2. *What will happen if only Alice and Bob were muddy?*
3. *What will happen if only Bob and Caroline were muddy?*
4. *What will happen if all three were muddy?*

Puzzle 17 (Lying) *From three children, Alice and Bob are muddy and Caroline is clean. Alice is hungry and does not want to play a game. She steps forward immediately at father's first clapping of hands, without considering if she knows whether she is muddy.*

1. *What does Bob conclude?*
2. *What does Caroline conclude?*
3. *What does Alice conclude, if she were after all thinking about what she did?*

Now suppose all three children are muddy, and that again Alice steps forward after the first clapping of hands. What will Bob and Caroline now conclude?

Some versions of the muddy children puzzle are not about mud on your forehead, but about hats or caps on your head. (In India a version goes round where you do not know the color of your eyes.) Just as you cannot see whether you have mud on your forehead, you also cannot see what the color is of the hat on your head. For the hats version (it exists for children, but also for prisoners, or wise men), there is a version where the children are not standing in a circle (so that they can all see each other), but where they are standing in a line and are all facing the same direction. Every child can see the hats on the heads ahead of it, but not its own hat and also not the hats of the children behind it. In order to have clear vision, they had better queue up on a staircase, such that each child further ahead is standing slightly lower. (There is a version of four people standing in line, for the Dalton Brothers from the *Lucky Luke* comics series. They do not need a staircase, as they are all of different height.)

Puzzle 18 *Ten children stand in line all facing the same direction, with a white hat or a black hat on their head. From the hindmost child forward, they are all allowed to guess (aloud!) the color of their hat: black or white. It is possible that all but one child makes a correct guess. How can they manage to do that? Before they are going to stand in line and have the hats placed on their heads, they are allowed to agree on a protocol.*

The last version of muddy children is from the well-known puzzlebook *The Princess and the Tiger* by Smullyan (1982).

Puzzle 19

> *Three subjects—A, B, and C—were all perfect logicians. Each could instantly deduce all consequences of any set of premises. Also, each was aware that each of the others was a perfect logician. The three were shown seven stamps: two red ones, two yellow ones, and three green ones. They were then blindfolded, and a stamp was pasted on each of their foreheads; the remaining four stamps were placed in a drawer. When the blindfolds were removed, A was asked, "Do you know one color that you definitely do not have?" A replied, "No." Then B was asked the same question and replied, "No."*
>
> *Is it possible, from this information, to deduce the color of A's stamp, or of B's, or of C's? (Smullyan 1982, p. 6—7)*

3.4 History

An old source for the muddy children puzzle is a German translation of the French literary classic *Gargantua et Pantagruel* (Rabelais, sixteenth century). This well-known work has been translated into many languages. This German translation by Regis (1832) was discussed by Born et al. (2008). It contains an extensive notes section, added by the translator, that contains an entry on the phrase "Ungelacht pfetz ich dich." (In French, this is "pincer sans rire." It relates to a game called La Barbichette. The game is also being played in the first volume of the Asterix comics series.)

> Ungelacht pfetz ich dich. Gesellschaftsspiel. Jeder zwickt seinen rechten Nachbar an Kinn oder Nase; wenn er lacht, giebt er ein Pfand. Zwei von der Gesellschaft sind nämlich im Complot und haben einen verkohlten Korkstöpsel, woran sie sich die Finger, und mithin denen, die sie zupfen, die Gesichter schwärzen. Diese werden nun um so lächerlicher, weil jeder glaubt, man lache über den anderen.
>
> I pinch you without laughing. Parlor game. Everybody pinches his right neighbor into chin or nose; if one laughs, one must give a pledge. Two in the round have secretly blackened their fingers on a charred piece of cork, and hence will blacken the faces of their neighbors. These neighbors make a fool of themselves, since they both think that everybody is laughing about the other one.

Instead of mud on foreheads, here we have charcoal on noses or chins. You cannot see if your own nose has been blackened. A crucial difference is the absence of synchronization. (When does you neighbor start laughing? There is no clapping of hands.) We have no record of publications involving the riddle between 1832 and the 1950s. Surely that is for lack of consulting the proper references in the proper language! The riddle makes its reappearance only slightly later than occurrences of other epistemic riddles from the 1940s onward (e.g., on ignorance about ages or house numbers) in magazines such as *Strand Magazine*.

In 1953, the muddy children puzzle for three children, for the case where they are all muddy, appears in *A Mathematician's Miscellany* by Littlewood (1953), right at the beginning. (This book is also an original source for the consecutive number riddle.) It is as follows:

> Three ladies, *A*, *B*, *C* in a railway carriage all have dirty faces and are all laughing. It suddenly flashes on *A*: why doesn't *B* realize *C* is laughing at her?—Heavens, I must be laughable. (...) (Littlewood 1953, p. 3–4)

Interestingly enough, Littlewood calls (solving) the puzzle a typical example of nontrivial mathematical reasoning. In the solution, he treats the case where all children are muddy. (The puzzle seems a bit of tongue-in-cheek to his colleague Hardy, who in (Hardy 1940) cites Euclid's proof that there are infinitely many primes as a typical example of nontrivial mathematical reasoning; a classical mathematical problem, not a modern one.)

A version of muddy children with hats instead of mud is found in the weekly magazine *Katholieke Illustratie,* by van Tilburg (1956). This is one of 626 such "breinbrouwsels" (brain crackers) that appeared in that magazine between 1954 and 1965, namely Breinbrouwsel 137 (volume 90, issue 32, page 47) entitled "Doe wel en zie niet om" (Do well and don't look back). Instead of mud, we now have colored caps worn by a team of four rowers sitting in front of one another, where a rower can only see the color of the caps in front of him, but not the color of his own cap or that of those behind him. Hans van Ditmarsch and Rineke Verbrugge spent several days in the Groningen University Library leafing through decades of volumes of this magazine *Katholieke Illustratie,* uncovering this information. They were actually looking for an older version of the sum-and-product riddle of Chapter 7, which they did not find.

Another source from the 1950s is a puzzlebook by Gamow and Stern (1958). This sketches a situation with 40 unfaithful women. Everyone knows if a woman is unfaithful, except her own husband. A version with unfaithful men is found in Moses et al. (1986; this also contains many other versions of the riddle for synchronous or asynchronous conditions); McCarthy (1978) publishes a politically more correct version of the puzzle, namely for "wise men." The wise men have to find out the color of the dot on their forehead.

The currently most popular version of the riddle, with muddy children, was first presented by Barwise (1981). Strangely enough, this by now best-known setting of the puzzle is much like the 1832 setting. As far as we know, this is a coincidence, and Barwise was not aware of the older version.

4

Monty Hall

Suppose you have made it to the final round of a game show. You can win a car that is behind one of three doors. The game show host asks you to pick a door. You choose door number 1. He tells you that he knows where the car is, and opens one of the remaining doors that does not have the car, say door number 3. Now he asks you whether you would like to switch to door number 2. Should you switch doors?

This puzzle often leads to furious debates. A lot of people have very strong intuitions about it: most think it does not matter if you switch or not. There are two remaining doors that might contain the car; you do not know which one, so it seems reasonable to assume that both doors are equally likely to contain the car. Hence, it should not really matter whether you switch. Surprisingly, it *does* matter whether you switch. Before we discuss this, we will first test our intuitions about probabilities.

4.1 What Is the Best Question to Ask?

Puzzle 20 *Anthony and Barbara play the following game. First, Barbara selects a card from an ordinary set of 52 playing cards. Then, Anthony guesses which card Barbara selected. If he guesses correctly, Barbara pays him 100 euros. If he guesses incorrectly, Anthony pays Barbara 4 euros. To make the game a bit more fair, Anthony is allowed to ask a yes/no question before he guesses, and Barbara has to answer his question truthfully. Which question is better: "Do you have a red card?" or "Do you have the Queen of hearts?"*

It almost seems obvious that it would be better for Anthony if he were to ask the first question. After this question is answered it is guaranteed that half of the cards can be excluded by Anthony, whereas it is very likely that after the second question only one card can be excluded. Yet, this intuition is wrong.

In fact, it does not matter which question Anthony asks. Both questions give equally good chances for Anthony to win the game. Let us assume that Barbara has taken a card at random and so the probability for each card is the same, namely $\frac{1}{52}$. When Barbara answers the question "Do you have a red card" there are only 26 possibilities left for Anthony (regardless of whether her answer is "yes" or "no"). The probability that Anthony guesses her card correctly now is $\frac{1}{26}$.

When Anthony asks Barbara whether she has the Queen of hearts and her answer is "yes," then Anthony will obviously correctly guess which card she has. The probability that this occurs is only $\frac{1}{52}$. The probability that she answers "no" is $\frac{51}{52}$. Subsequently, Anthony will guess correctly with probability of only $\frac{1}{51}$. In total, this equals $\frac{51}{52} \cdot \frac{1}{51} = \frac{1}{52}$. The probability that Anthony guesses correctly after asking the second question, equals the probability of him guessing correctly when he gets the answer "yes" plus the probability of him guessing correctly when he gets the answer "no." This is $\frac{1}{52} + \frac{1}{52} = \frac{1}{26}$. This is the same probability of guessing correctly as when Anthony asks the first question. So, it does not matter which question he asks.

Puzzle 21 *Anthony and Barbara play the same game, but now Anthony can ask a question with four possible answers. Which question is better: "What do you have: clubs, hearts, diamonds or spades?" or "What do you have: the Queen of hearts, the Three of diamonds, the Ace of spades, or another card?"*

In this case too it does not matter which question Anthony asks. Both questions make it equally probable for Anthony to guess Barbara's card. For the first, only 13 cards remain after the question is answered. So, the probability that Anthony guesses correctly is $\frac{1}{13}$. For the other question the probability of guessing correctly is $\frac{1}{52} + \frac{1}{52} + \frac{1}{52} + \frac{49}{52} \cdot \frac{1}{49} = \frac{1}{13}$. So the probabilities are the same.

The only thing that matters in these games is the number of answers a question has. When the question has only one possible answer the probability to win is $\frac{1}{52}$, because you do not learn anything from its answer. A question with one possible answer would be "Do you have a card?", to which the answer is "Yes." (This is not really much of a question.) When there are two possible answers, the probability is $\frac{2}{52}$ to win, as we learned from Puzzle 20. With three possible answers, the probability is $\frac{3}{52}$, and with four possible answers (such as in Puzzle 21) it is $\frac{4}{52}$, and so on. When there are 52 possible answers, the probability of guessing correctly is $\frac{52}{52} = 1$. You can check this yourself. The most obvious question with 52 possible answers is "Which card do you have?" After learning the answer, you are guaranteed to guess my card correctly.

The previous two puzzles showed that our intuitions can be quite misleading when it comes to probability. A simple calculation can put us back on the right track. This seems rather difficult in the puzzle about the game show host and the car.

4.2   Why Is It to Your Advantage to Switch Doors?

It is to your advantage to switch doors, because the probability that you win the car by switching is $\frac{2}{3}$. In solving this puzzle we make some assumptions. First of all, we assume that the car has been placed behind one of the doors randomly. Initially, the probability that the car is behind door number 1 is $\frac{1}{3}$ and the same holds for door number 2 and door number 3. Furthermore, we assume that the rules of the game show are such, that the host opens a door that you did not pick initially and which does not contain the car. Lastly, we assume that if the host can choose between opening two doors, he does so randomly.

Suppose that you initially pick the door with the car. The probability of this is $\frac{1}{3}$. In that case you will not win the car by switching.

Suppose that you initially do not pick the door with a car. The probability of this is $\frac{2}{3}$. Now the host has to open a door. There is only one door the host is allowed to open according to the rules. You picked a door, and he cannot open that door. The car is behind another door, and he cannot open that door either. So the host opens the remaining door. If you switch to the closed door, you automatically end up with the door that has the car behind it.

The probability of winning the car by switching doors is therefore $\frac{1}{3} \cdot 0 + \frac{2}{3} \cdot 1 = \frac{2}{3}$.

A lot of people are not convinced by this argument. The best way of convincing yourself that it really leads to the right conclusion, is by playing the game yourself with three teacups. You have to act the part of the game show host as well as the part of the candidate. Play it 50 times and record how often the candidate wins the car. You can use a die to determine where the car is to be placed; for example, when you throw 1 or 2, you place the car behind door number 1, when you throw 3 or 4, you place the car behind door number 2 and when you throw 5 or 6, you place the car behind door number 3. You can also use the die to determine which door the candidate picks initially, and (if there is any choice) which door the host opens. If you let the candidate switch doors every game, it is likely that he wins the car in more than 30 and less than 40 games, and it is unlikely that he wins the game in more than 10 and less than 20 games. Of course every other outcome is also possible, even not ever winning the car at all. But such outcomes are very improbable. If that happens, just play it another 50 times to convince yourself.

4.3 Versions

Some can be convinced that the solution of the puzzle is correct by considering a version with many more doors. Consider the following variation.

Puzzle 22 *Suppose you have made it to the final round of a game show. You can win a car that is behind one of a thousand doors. The game show host asks you to pick a door. You choose door number 1. He tells you that he knows where the car is and opens all of the remaining doors except door number 1 and door number 899. Now he asks you whether you would like to switch to door number 899. Should you switch doors?*

If you are still not convinced that switching doors is the best course of action, and playing the game yourself 50 times did not convince you either, then we have little more to offer to convince you.

As we indicated, one of the assumptions we made, was that the host randomly opens a door. But the host could behave in very different ways.

Puzzle 23 *Suppose we know something more about the game show host: he is incredibly lazy. He does not want to waste his energy and during the show he prefers to walk as little as possible in the studio. Let us assume that door number 1 is closer to the host than door number 2 and that door number 2 is closer to him than door number 3. Suppose again that you initially choose door number 1. Now the host has to open one of the other doors that does not contain the car. He walks all the way to door number 3 and opens it. Should you switch doors? (And what is the probability of winning by switching in this case?)*

Puzzle 24 *Take the same scenario as in puzzle 23, but now assume that the host is very athletic and likes to walk as much as possible. You choose door number 1 and the host opens door number 3. Should you switch doors? (And what is the probability of winning by switching in this case?)*

As we see, the host plays a very important role in the puzzle. Now let us consider a version of the puzzle where he plays no role whatsoever. Would that make a difference?

Puzzle 25 *Suppose you have made it to the final round of a game show. You can win a car that is behind one of three doors. The game show host asks you to pick a door. You choose door number 1. He tells you that he knows where the car is, but before he can do anything, door number 3 opens due to a technical error. There is no car behind it. The host offers you the opportunity to switch to door number 2. Should you switch doors?*

4.4 History

This puzzle is best known as "The Monty Hall Dilemma" after the famous American game show host Monty Hall. The puzzle was first associated with Monty Hall by Selvin (1975b, a). Versions of the puzzle were already circulating in the 1960s (for example Mosteller (1965)), but these were about three prisoners where one of them was about to be hanged. In the 1990s, interest in the puzzle was sparked by a discussion of the problem by vos Savant (1990), where the doors without a car contained goats. This led to a heated debate in the USA.

The card game in Puzzle 20 was discussed in the dissertation of Kooi (2003), where he argues that by analogy, in the game of Mastermind one should always ask the question that has the most possible answers.

5
Russian Cards

From a pack of seven known cards 0, 1, 2, 3, 4, 5, 6 *Alice and Bob each draw three cards and Cath gets the remaining card. All this is known. How can Alice and Bob openly inform each other about their cards, without Cath learning of any of their cards who holds it?*

"All this is known" means that the players know that there are seven cards, that they know how many cards the other players have, and that all players only know their own cards; but it also means that they know that the other players know this, and so on. This is as in the other riddles."Openly inform each other" means that they have to talk aloud: If Alice is saying something, then Bob and Cath will hear it, and if Bob is saying something, then Alice and Cath will hear that. "Openly" also means that Alice and Bob must not show their cards to each other, or in some way or the other get a peek at them. Otherwise, that would be a very simple way to get to know the card deal without Cath knowing it! The situation is, therefore, as in a real card game: All actions have to be public, otherwise the game is unfair. We further assume that the players only tell the truth. As both Alice and Bob have an interest in finding out the truth about each others' cards, that seems a reasonable restriction.

Suppose Alice holds the cards 0, 1, and 2; Bob holds the cards 3, 4, and 5; and Cath holds the card 6. Instead of saying (and writing) that Alice holds the set {0, 1, 2}, we will say that Alice's hand of cards is 012, and similarly that Bob's hand of cards is 345, and we will write 012.345.6 for that card deal. To simplify the exposition, this will always be the actual card deal.

5.1 You Had Better Know What You Say

A first attempt is as follows:

Alice says to Bob: "You have 012 or I have 012," after which Bob says to Alice: "You have 345 or I have 345."

This may appear to be a solution, but it is not a solution. How come? The real card deal is 012.345.6. First, Alice says to Bob, "You have 012 or I have 012." There are four card deals wherein Alice has 012, namely

> 012.345.6
> 012.346.5
> 012.356.4
> 012.456.3

and there are four card deals wherein Bob has 012, namely

> 345.012.6
> 346.012.5
> 356.012.4
> 456.012.3

These eight card deals seem to be still possible after Alice says to Bob, "You have 012 or I have 012." Now it is Bob's turn. Bob says to Alice, "You have 345 or I have 345." Bob has 345 or Alice has 345 in two of the eight card deals above:

> 012.345.6
> 345.012.6

Therefore, those two card deals seem to remain possible after Bob's announcement. Cath has card 6, and therefore does not know from any other card whether it is held by Alice or Bob. For example, Alice holds card 0 in the actual card deal 012.345.6, Bob holds card 0 if the card deal had been 345.012.6. And so on, for all other cards except 6.

Still, this is not a solution. We can realize why, if we take into consideration the knowledge of the players about other possible card deals. There are 140 different deals of 7 cards over 3 players where 2 of the 3 players get 3 cards each.

Alice draws three cards out of seven. For the first card there are 7 options, for the second card 6 remain, and for the third, 5. Of course it does not matter if she draws card 0 as the first of her three cards, as the second of her three cards, or as the third of her three cards. So we have to take such identifications into account. Take any of her three cards. This card could have been drawn first, second, or third, either of 3 ways. And a card from the remaining two cards could have been the second or the third draw, one of 2. Altogether, we therefore have $\frac{7 \cdot 6 \cdot 5}{3 \cdot 2} = 35$ ways to draw three out of seven cards. (We computed $\binom{7}{3} = 35$.)

Then, Bob draws three out of the remaining four cards. As there remains only one card after that, the number of different draws is the same as the number of remaining cards. There are 4 possible remaining cards!

Finally, Cath has no choice. She just gets the remaining card.

Altogether we, therefore, have $35 \cdot 4 \cdot 1 = 140$ card deals.

Whatever Alice's hand of cards is, in the initial situation she cannot distinguish four card deals from one another. This is because there are four cards she does not hold, and she considers it possible that Cath holds any of those cards. Given that the actual card deal is 012.345.6, she cannot distinguish card deals:

$$012.345.6$$
$$012.346.5$$
$$012.356.4$$
$$012.456.3$$

For Bob, a similar story holds. He also considers four card deals possible, but not the same card deals as Alice, namely

$$012.345.6$$
$$016.345.2$$
$$026.345.1$$
$$126.345.0$$

Cath considers many more card deals possible, namely 20: For any given card held by Cath, she does not know which triple has been drawn by Alice from the remaining six cards; and there are $\frac{6 \cdot 5 \cdot 4}{3 \cdot 2} = 20$ ways to choose three cards out of six. We will not list the 20 card deals that are possible, given her card 6.

If the players make announcements about their cards, this total of 140 possible card deals will be reduced.

As we said, we assume that the actual card deal is 012.345.6, and we suggested that Alice's announcement "You have 012 or I have 012" reduced the number of possible card deals to eight, and that subsequent to Bob's announcement "You have 345 or I have 345" two possible card deals remain. We now introduce models to represent these three different information states of the game and the uncertainty of the players about each others' cards. For the initial 140 card deals, this is a bit cumbersome, so we only do that schematically. But for the other two information states, we are explicit: First we get an information state with eight card deals, then we get one with two card deals.

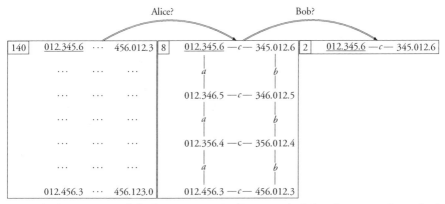

Links between card deals represent indistinguishability between these deals, for the player whose name labels the link. For example, Cath cannot distinguish 012.345.6 from 345.012.6. Therefore, there is a c link in the middle picture above, between those two card deals. Alice cannot distinguish 012.345.6 from 012.346.5. Therefore, there is an a link between those two card deals. But she also cannot distinguish 012.345.6 from 012.356.4. Therefore, in the figure we can reach 012.356.4 from 012.345.6 by a *path* consisting of two such a links. Card deals connected by a path of links all labeled with a certain player's name cannot be distinguished from one another by that player.

Let us now analyze the announcements. Alice said, "You have 012 or I have 012." This is true, because she has 012. But how does Alice know that what she says is true? We put ourselves in the perspective of Cath:

Either Alice has 012, or she does not have 012.

Suppose Alice has 012. Her announcement "You have 012 or I have 012" is then true, because "I have 012" is true, and because "You have 012 or I have 012 " follows from "I have 012." No problem there!

Suppose Alice does *not* have 012. For "You have 012 or I have 012" to be true, Bob must then have 012, in which case Alice has one of 345, 346, 356, and 456. Suppose she has 345. How can she know in that case that the announcement "You have 012 or I have 012" is true? As she has 345, she considers it possible that Bob has 012 and that I have 6, but she also considers it possible that Bob has 016 and that I have 2. But in the last case the statement "You have 012 or I have 012" would be false! Now in the initial situation Alice cannot distinguish the card deals 345.012.6 and 345.016.2 from one another. Therefore, if she were to have 345 she would then not know that her announcement is true. As she may only say what she knows to be true, she will therefore not make that announcement if she has 345. But she made the announcement . . . .

Therefore, Alice must hold 012.

And therefore, Cath knows the entire deal of cards after Alice's announcement. Obviously, Bob also knows the card deal after Alice's announcement. Only Alice does not know the card deal yet: She cannot learn to distinguish the four card deals wherein she holds 012 by her own announcement. We conclude that after Alice's announcement, the number of possible card deals should be reduced to four and not to eight. But that also makes it rather immediately clear that any further announcement after Alice's announcement does not lead toward a solution. Cath already knows the card deal after her announcement! Bob's subsequent announcement "You have 345 or I have 345" only leads to a further reduction of possible card deals, namely from the four remaining possible card deals to the unique actual card deal 012.345.6.

The real consequences of Alice's and Bob's announcement are therefore:

			Alice			Bob	
140	012.345.6	⋯	456.012.3	4	012.345.6	1	012.345.6
	⋯	⋯	⋯		a		
	⋯	⋯	⋯		012.346.5		
	⋯	⋯	⋯		a		
	⋯	⋯	⋯		012.356.4		
	⋯	⋯	⋯		a		
	012.456.3	⋯	456.123.0		012.456.3		

It is important in this analysis that Alice only says what she *knows* to be true. Otherwise, how is she supposed to know that she is not lying? Saying what you know to be true is more restricted than merely saying what is true. You may by chance say something that is true without knowing it.

By introducing a fourth player Dirk, we can make the difference clear between saying what you know to be true and saying what is true. Imagine Alice, Bob, and Cath sitting around a table, and the insider Dirk who can walk around the table and look into everybody's cards, with permission of all players and in full sight of them (so that there is common knowledge again of this matter). Now imagine the following conversation:

Dirk says: "Bob has 012 or Alice has 012," after which Dirk says: "Bob has 345 or Alice has 345."

This is not exactly the same as what Alice and Bob said:

> Alice says to Bob: "You have 012 or I have 012," after which Bob says to Alice: "You have 345 or I have 345."

But what Dirk says is identical to what Alice and Bob said if we rephrase their announcements as follows:

> Alice says: "Bob has 012 or Alice has 012," after which Bob says: "Bob has 345 or Alice has 345."

Alice, Bob, and Dirk only say what they know. But Dirk knows a lot *more* than Alice and Bob. Strangely enough, Dirk's announcements are therefore a lot *less* informative than Alice's and Bob's identical announcements. If Alice does not have 012, then Alice does not know that her announcement is true, because she cannot see Bob's cards, and therefore cannot know that Bob has 012. But if Alice does not have 012, then Dirk knows that his announcement is true, because he sees that Bob has 012, and therefore knows that Bob has 012.

It may be confusing that the same announcement can be interpreted in different ways depending on who is saying it. We can get around this by having the players "really" saying that they know that their announcement is true. Alice is not merely saying, "You have 012 or I have 012," but she is really saying, "I know that you have 012 or that I have 012." In other words, we compare the following two announcements:

> Alice knows that Bob has 012 or Alice has 012,
> Dirk knows that Bob has 012 or Alice has 012.

The first holds for four card deals, and the second holds for eight card deals.

The role of Dirk is similar to that of the quizmaster or the narrator in other riddles. He is the guy, or in some other riddles the girl, who knows everything about the setting of the riddle, and who knows it correctly. Because of that, the meaning of "Dirk knows that Bob has 012 or Alice has 012" is the same as the meaning of "Bob has 012 or Alice has 012." So we can also say that we are comparing "Alice knows that Bob has 012 or Alice has 012" with "Bob has 012 or Alice has 012," and in general, the meaning of any proposition with the more restricted meaning of "Player x knows the proposition."

5.2   Knowing What Another Player Knows

The following also does not solve the problem:

Alice says: "I don't have card 6" and Bob says: "I also don't have card 6."

Because Cath has card 6, Cath does not appear to learn from Alice's announcement what Alice's cards are, whereas Bob, who has 345, learns that Alice has 012. After that, Alice learns from Bob's announcement that he has 345, and Cath still does not seem to get any wiser. Again, this does not provide a solution. The problem is that Alice does not know that Cath has card 6 when she makes her announcement. Because Alice has 012, she can imagine that Cath has card 5. And in that case, Cath would learn from Alice's announcement that Bob has card 6. Therefore, Alice will not say that she does not have card 6.

Again, we could imagine an insider Dirk, who knows the card deal, to make both announcements. To be precise:

Dirk says: "Alice doesn't have card 6," and Dirk says: "Bob also doesn't have card 6."

And after those announcements Cath would still be ignorant. After the first announcement, the 20 card deals remain wherein Alice does not have card 6 (Alice can have three out of six cards), and after the second announcement the two remaining card deals are 012.345.6 and 345.012.6, so that Cath remains ignorant.

We can observe that:

* After the announcement "Alice doesn't have card 6," Cath does not know any of the other players' cards.
* After the announcement "Alice doesn't have card 6," Alice does not know that Cath does not know any of the other players' cards.

Because Alice does not know that Cath remains ignorant, she will not make the announcement.

If we see the above announcement of not having 6 as an execution of the protocol wherein Alice announces a card she does not hold, the reason that the protocol does not work seems to be that other executions do not preserve Cath's ignorance, such as announcing that she does not have 5. This suggests that a protocol may work if all its executions preserve Cath's ignorance. But such a protocol is also unsafe. Consider the following announcements:

Alice says: "I have 012 or I have none of those cards," and Bob says: "I have 345 or I have none of those cards."

There is only one execution of the (imaginary) protocol behind the announcement: *Say that you have your actual cards or none of those cards*. Again, if Dirk were to have made the announcements "Alice has 012 or Alice has none of those cards," and "Bob has 345 or Bob has none of those cards," then afterwards at least deals 012.345.6 and 345.012.6 would have remained so that Cath remains ignorant (of the ownership of any card except 6). The problem is when Alice and Bob say it. And this complication is rather interesting when described in terms of knowledge and ignorance. Consider the first announcement. The following deals of cards are still possible after Dirk says, "Alice has 012 of Alice has none of those cards."

$$
\begin{array}{ccccccc}
012.345.6 & —a— & 012.346.5 & —a— & 012.356.4 & —a— & 012.456.3 \\
| & & | & & | & & | \\
c & & c & & c & & c \\
| & & | & & | & & | \\
345.012.6 & —b— & 346.012.5 & —b— & 356.012.4 & —b— & 456.012.3 \\
| & & | & & | & & | \\
a & & a & & a & & a \\
| & & | & & | & & | \\
345.016.2 & —c— & 346.015.2 & —c— & 356.014.2 & —c— & 456.013.2 \\
| & & | & & | & & | \\
a & & a & & a & & a \\
| & & | & & | & & | \\
345.026.1 & —c— & 346.025.1 & —c— & 356.024.1 & —c— & 456.023.1 \\
| & & | & & | & & | \\
a & & a & & a & & a \\
| & & | & & | & & | \\
345.126.0 & —c— & 346.125.0 & —c— & 356.124.0 & —c— & 456.123.0 \\
\end{array}
$$

We recall that the actual deal of cards is 012.345.6. In that case, Cath indeed does not know any of the other players' cards, as she cannot distinguish card deals 012.345.6 and 345.012.6. But there is much more to say about this model, which will explain why it results from Dirk's announcement but not from Alice's similar announcement.

We also have (unlike before) that Alice knows that Cath does not know any of the other players' cards. Alice cannot distinguish deals 012.345.6, 012.346.5, 012.356.4, and 012.456.3. In all those cases, Cath does not know any of the other players' cards, for (another) example, if the deal is 012.346.5, then Cath cannot distinguish it from 346.012.5, and if the deal is 012.356.4, then Cath cannot distinguish it from 356.012.4, and similarly for the last possibility.

But Cath does not know *that*, i.e., Cath does not know that Alice knows that Cath does not know any of the other players' cards. Cath cannot distinguish card deals 012.345.6 and 345.012.6. If 345.012.6 had been the actual deal, then Alice would have been unable to distinguish it from 345.016.2, wherein

Cath knows that Bob has cards 0 and 1 (Bob has cards 0 and 1 in all four card deals that Cath cannot distinguish: 345.016.2, 346.015.2, 356.014.2, and 456.013.2). So Alice would not have made her announcement if her hand of cards had been 345. Similarly, when Alice would have held 346, 356, or 456.

So, Cath can conclude that Alice would only have made her announcement if she holds 012. Therefore, Cath knows all of Alice's cards.

Summing up, after the announcement "Alice has 012 or Alice has none of those cards":

* Cath does not know any of the other players' cards.
* Alice knows that Cath does not know any of the other players' cards.
* Cath does not know that Alice knows that Cath does not know any of the other players' cards.

Cath may assume that Alice knows that Cath does not know any of the other players' cards after her announcement. Therefore, if Alice makes the announcement and not Dirk, Cath can restrict the above model to the four card deals (the first row), wherein it is true that Alice knows that Cath does not know any of the other players' cards after her announcement. And therefore Cath learns the entire card deal.

Analyzing Bob's subsequent announcement has no additional value, as Cath already learns the entire card deal from Alice's announcement.

5.3   Solution of the Problem

By now, we have had various examples wherein Alice and Bob each make an announcement, but still something goes wrong. It is not so clear how one can systematically search for a solution to the riddle, because according to the rules of the game, Alice and Bob may say something more than once, and Alice and Bob may say just about anything as long as they say it openly. There is a wealth of options here! Alice says to Bob, "I have card 3, or if I have card 2 then you have 6 or Cath has 5, or if I have any of the cards 0, 4, or 6 then Cath has card 4 if you have card 2." It is nontrivial to investigate the informative consequences of such an elaborately structured proposition. Fortunately, things are simpler than that. The initial state of information is a well-described, finite model, and any informative announcement, as it is public, results in a restriction of that finite structure. There are only finitely many restrictions. But it gets even better: We only have to consider announcements of a special form. Anything a player can say is equivalent to an announcement, wherein that player says that its hand is one of a set of alternatives, the set of alternatives should, obviously, include the actual hand of cards of that player, as players are only allowed to say the truth, and what they know to be the truth. It may sound good that we can restrict our search for the solution that way, but note that the number of announcements is exponential in the number of alternative hands of cards of the announcing player. (There are 35 different hands of cards a player may have, and therefore 2^{35} different announcements to consider.) Announcements such as "I do not have card 6" or "Cath has card 6" can always be replaced by (for Alice) "I have one of 012 or . . . " and (for Bob) "I have one of 345 or . . . ." How many alternatives should be given for the actual hand of cards? One? Two? More than two? Because players can reason about each other's knowledge, a requirement is that there is a sufficient number of alternatives such that, no matter what any player's actual cards are, Cath would not learn any cards of Alice or Bob.

Suppose Alice says, "I have 012 or I have 134." What can go wrong? If Cath has 3, then she learns from this that Alice has 0, 1, and 2. If Alice were to have 134 and Cath had 0, then she learns that Alice has 1, 3, and 4. Always when Alice names two hands only, at least one of the cards occurring in these two hands will not be Alice's own. But if Cath were to have that card, she would learn all of Alice's cards. So, two hands are not enough.

Now suppose Alice says, "I have 012, 034, or 156." This remains a dangerous thing to say. Cath has 6, and therefore learns that Alice has 0. For other combinations of three hands, we run into similar trouble.

Even four hands of cards are not enough for Alice to know that her announcement does not leak information to Cath. The following argument explains this:

Four triples contain $4 \cdot 3 = 12$ card occurrences. As there are seven cards, at least two of those must occur only once. We can also assume that this is exactly two. Because if it had been more than two cards, there must be yet another card occurring three times. If Cath had that card, then she would have been able to eliminate all but one triple—that must therefore be the actual hand of Alice. And thus Cath would have learnt the entire card deal.

Choose a triple that contains a card i that occurs only once in the set of four triples. At least one of the other two cards in that triple must occur twice, let us say card j. (Otherwise there would have been three cards occurring only once in the four triples.) Now suppose Cath has that card. The two remaining triples that do not contain j also do not contain i. But that means that Cath would then learn that Bob must have card i. Lost again . . . .

Any solution must therefore consist of an announcement by Alice that contains at least five triples (hands). Indeed, there is such a solution:

> Alice says: "I have one of 012, 034, 056, 135, and 246," after which Bob says: "Cath has card 6."

Bob has 345 and learns from Alice's announcement that Alice has 012: One or more of Bob's cards occur in any of these five triples except 012. Whatever Alice's actual hand of cards, Bob would have learnt Alice's card from the announcement. For example, if Alice had 246, then Bob may have one of the following hands: 013, 035, 135, and 015. If it had been 013, then one of 0, 1, and 3 occurs in all hands of Alice's announcement except 246. So Bob learns Alice's hand of cards. Similarly for the other three hands of Bob in case Alice has 246. And so on, for the other hands in Alice's announcement.

We also have to show that no matter Alice's actual hand of cards, Cath would not have learnt any of Alice's or Bob's cards.

Suppose Cath had card 0. Then Alice could have had 135 or 246. Cath now does not learn any of Alice's or Bob's cards, because each of the numbers 1, 2, 3, 4, 5, and 6 occurs in at least one of these two hands (so Cath cannot conclude that Bob has it), and each of these numbers also is absent in at least one of these two hands (so Cath cannot conclude that Alice has it).

Suppose Cath had card 1. Then Alice could have had 034, 056, or 246. Each of 0, 2, 3, 4, 5, and 6 occurs at least once in one of these three hands and also is absent at least once.

Etcetera for all other cards Cath may hold.

Schematically, we can visualize the result of Alice's announcement as follows, where in this case we chose a slightly simpler visualization without links labeled with player names. Below, the card deals in a row cannot be distinguished by Alice and the card deals in a column cannot be distinguished by Cath. Bob can distinguish all card deals.

012.345.6	012.346.5	012.356.4	012.456.3			
034.125.6	034.126.5			034.156.2	034.256.1	
		056.123.4	056.124.3	056.134.2	056.234.1	
135.024.6		135.026.4		135.046.2		135.246.0
	246.013.5		246.015.3		246.035.1	246.135.0

We now analyze Bob's announcement "Cath has card 6." Bob can say this truthfully, because he knows the card deal. Alice obviously learns from this what Bob's cards are. Bob's announcement is just as informative as an announcement listing alternatives for his (Bob's) actual hand of cards:

> Bob says: "I have one of 345, 125, and 024."

From the 20 deals above, the following 3 now remain (namely those in the column where Cath has card 6). Clearly, Cath remains ignorant.

012.345.6
034.125.6
135.024.6

And with that verification, we have finally settled that this is a solution of the Russian cards problem. There are also other solutions for the Russian cards problem. For other numbers of cards per player, there are yet other solutions. And for more than three players, there are even other solutions. These problems are truly in the intersection of the logic of knowledge and combinatorial mathematics. A solution for such a generalized Russian cards problem is a protocol consisting of a set of sequences of announcements that can be alternatingly made by Alice and Bob, where what they say can always be seen as a collection of alternatives for their actual hand of cards. After such a sequence, that can be seen as an execution of such an underlying protocol, Alice should know Bob's cards and Bob should know Alice's cards, but Cath should not learn any of Alice's or Bob's cards, no matter the deal of cards for which the announcements could have been truthfully made.

We close this chapter with other solutions for the Russian cards problem and some other variations and generalizations.

5.4 Versions

Puzzle 26 *Suppose that:*

Alice says: "I have one of 012, 034, 056, 135, 146, 236, 245." *and after that Bob says: "Cath has card 6."*

Show that this is a solution, and that this solution is different from the solution wherein Alice announces five alternative hands of cards.

In the next puzzle, we use that, given natural numbers x and y, "x modulo y" is the remainder of x after division by y.

Puzzle 27 *Suppose that:*

Alice announces the sum modulo 7 of her cards, after which Bob announces Cath's card.

Demonstrate why this is a solution. Is it different from other solutions?

Puzzle 28 *Suppose that Cath may learn one of Alice's or Bob's cards, but just not all of them. Otherwise, nothing changes in the setting of the problem. There are now more succinct ways for Alice and Bob to inform each other about their hands of cards, without Cath learning the card deal. Give one such solution.*

In the next puzzle, we only consider Alice informing Bob. It should be clear by now that Alice's announcement consists of a number of alternatives for her hand of four cards. Assume that the actual deal of cards is 0123.456789A.BC, where A, B, C represent the numbers 10, 11, 12 (as in hexadecimal, base 16, counting).

Puzzle 29 *Alice, Bob, and Cath hold, respectively, 4, 7, and 2 cards. How can Alice openly inform Bob about her cards, without Cath learning of any of their cards who holds it?*

We finish with a puzzle for more than three players, and where the protocol consists of more than two steps. We also weaken the requirements: The eavesdropper may not learn for any card other than her own that it is held by another player; but the eavesdropper may learn for a card other than her own that it is *not* held by another player. For three players, if Cath knows that a card is not held by Alice, then it must be held by Bob. But for four players Alice, Bob, Cath and Eve, if Eve knows that a card is not held by Alice, then Eve may still be uncertain if it is held by Bob or by Cath.

To solve the next puzzle, it is convenient to model the announcements as statements about the ownership of individual cards (such as the announcement wherein Bob announces Cath's card), and not as alternatives of hands of cards.

Puzzle 30 *Alice, Bob, and Cath hold, respectively, 2, 3, and 4 cards. There is also a fourth player Eve, who is in this case the eavesdropper and wants to learn the card deal. Eve holds no cards. How can Alice, Bob and Cath openly inform each other about their cards, without Eve learning of any of their cards who holds it?*

5.5 History

The oldest known source for this riddle is the journal article *A Problem in Combinations* by Kirkman (1847)—a nineteenth century mathematician who was working in England. The problem was posed at the Mathematics Olympiad in Moscow in 2000. The jury was then confronted with solutions of the kind:

Alice says to Bob: "You have 012 or I have 012," after which Bob says to Alice: "You have 345 or I have 345," and found it difficult to disqualify such solutions. Alexander Shen was involved in that Mathematics Olympiad. This story was later reported in *The Importance of Being Formal* by Makarychev and Makarychev (2001). The modulo-sum approach of Puzzle 27 was the reported correct answer. Alexander Shen told Marc Pauly about this, and Marc Pauly told Hans van Ditmarsch. An outcome of those communications is the publication *The Russian Cards Problem* by van Ditmarsch (2003). This is the logical analysis presented in this chapter. At the time, Hans thought that the problem originated in Moscow. He therefore called it Russian Cards. Only later, he found out about the older source Kirkman. He then wanted to change the name of the riddle. But, by then, it was already out of his hands. The name had stuck in the community. It is still called Russian cards, and any further elaborations go under the name *generalized* Russian cards problem. A Dutch-language publication on the problem is van Ditmarsch (2002c).

The Russian cards problem is about seven cards. In the section "Versions," we saw some generalizations of this riddle for other deals of cards over three players, for other secrets than ownership of single cards, or for even more players. Puzzle 28 enters the area of protocols for secret bit exchange (Fischer and Wright 1992); Puzzle 29 is taken from Albert, Aldred, Atkinson, van Ditmarsch, and Handley (2005; it may have older roots too), announcing the sum of your cards modulo a prime number is treated more generally by Cordón-Franco et al. (2012), another recent treatment of card deal protocols is by Swanson and Stinson (2014); and Puzzle 30 uses the technique developed by Fernández-Duque and Goranko (2014).

6
Who Has the Sum?

Anne, Bill, and Cath all have a positive integer on their forehead. They can only see the foreheads of others. One of the numbers is the sum of the other two. All the previous is common knowledge. They now successively make the truthful announcements:

1. *Anne: "I don't know my number."*
2. *Bill: "I don't know my number."*
3. *Cath: "I don't know my number."*
4. *Anne: "I know my number. It is 50."*

What are the other numbers?

6.1 A Binary Tree of Uncertainty

Anne has 50. One of the three numbers is the sum of the other two. Anne's number must be the sum or the difference of the two numbers she sees—if it is the difference, then one of the numbers she sees must be the sum of her own number and the other number she sees. It is not so clear how this helps us solve the problem. There are infinitely many possibilities: she could see 16 and 34 (in which case she would not know if her number is 50 or 18), but she could equally well see 250 and 200 (in which case she would not know if her own number is 50 or 300), or 2 and 48 (50 or 46?), or . . . She could also see 25 twice. Their difference is 0, and 0 is not allowed. Only positive integers are allowed. She would then know that she has 50. But in her first announcement she says that she does not know. So she clearly does not see 25 twice. Well, that is a start. Let us now systematically investigate this matter.

We represent a situation by a triple, for example, $(5, 8, 13)$, where the first argument is Anne's number, the second Bill's number, and the third Cath's number. Anne cannot distinguish this triple from $(21, 8, 13)$. She is uncertain whether she has 5 or 21.

Whenever Anne sees two identical numbers, she knows that her own number must be their sum. She *knows,* because her number is the sum or the difference,

and as their difference is 0, and 0 is not allowed, it therefore has to be the sum. Let the triple be $(50, 25, 25)$. She sees 25 twice, so her own number must be 50. In that case, Bill and Cath do not know their own number. Bill cannot distinguish triple $(50, 25, 25)$ from triple $(50, 75, 25)$, whereas Cath cannot distinguish $(50, 25, 25)$ from $(50, 25, 75)$. And if it had been $(50, 75, 25)$, then Anne would not have been able to distinguish this from $(100, 75, 25)$, and Cath that one from $(50, 75, 125)$. And so on. We can visualize the agents' uncertainty about their number in a tree. The root of the tree is $(50, 25, 25)$. This comes on top. (Mathematical trees grow upside down.) The triples $(50, 75, 25)$ and $(50, 25, 75)$ come below it, where we indicate by a label for which agent the lower triple is indistinguishable from the root. Then, triples $(100, 75, 25)$ and $(50, 75, 125)$ come below $(50, 75, 25)$, etc. The general rule is that a triple is put *below* another triple if one of its arguments is bigger.

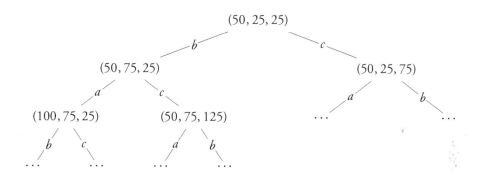

The general pattern is as follows. Nodes are triples (x, y, z) such that $x = y+z$ or $y = x + z$ or $z = x + y$. Anne's forehead contains the first argument. Therefore, from her perspective, the number x must be the sum of y and z or the difference between y and z, i.e., whichever is the larger one of the two minus the smaller one of the two. In the picture below, we therefore give the absolute value $|y - z|$ of that difference:

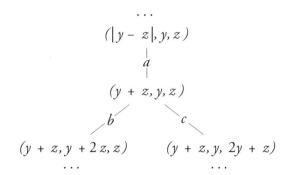

The tree with root $(50, 25, 25)$ is much like any tree with root $(2x, x, x)$, for $x \geq 1$. For example, take the tree with root $(10, 5, 5)$. In the $(50, 25, 25)$ tree, the root is indistinguishable for Bill from $(50, 75, 25)$, whereas in the $(10, 5, 5)$ tree the root is indistinguishable for Bill from $(10, 15, 5)$. We get the same shape of tree. The smallest numbers occur in the tree with root $(2, 1, 1)$. Knowledge analyses applied to the tree with root $(2, 1, 1)$ apply just as well to the tree with root $(10, 5, 5)$, or with root $(50, 75, 25)$, and in fact to any tree with root $(2x, x, x)$. The tree with root $(2, 1, 1)$ is as follows. For readability, we write 211 instead of $(2, 1, 1)$, and so on.

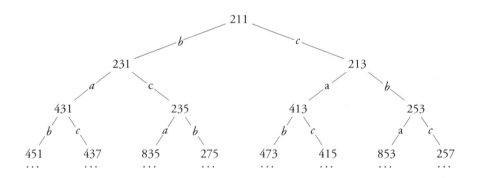

6.2 Informative Announcements

Let us now see what happens to this information structure, when processing the three successive announcements. The first announcement is:

* *Anne: "I don't know my number."*

We eliminate the state wherein Anne would have known this, namely $(2, 1, 1)$. The resulting structure is:

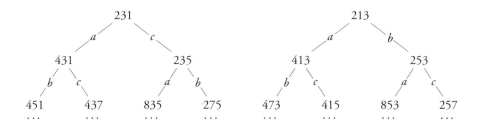

The second announcement is:

* *Bill: "I don't know my number."*

Again, we eliminate the states (triples) wherein Bill would have known this, as these are now no longer possible. And, of course, we do this in the situation resulting from processing Anne's announcement, not in the initial state of information. This means that we can remove the state $(2, 3, 1)$ but not the state $(2, 1, 3)$. How come? If the state is $(2, 3, 1)$, Bill was initially not able to distinguish this from state $(2, 1, 1)$—however, $(2, 1, 1)$ is ruled out because Anne said she did not know her number. Therefore, Bill then knows that his number is 3. But he said he did not know. Therefore, the triple is not $(2, 3, 1)$. State $(2, 1, 3)$ is indistinguishable for Bill from state $(2, 5, 1)$, lower down in the tree. Those triples therefore remain—as do all other triples in both trees, as also there Bill does not know his number.

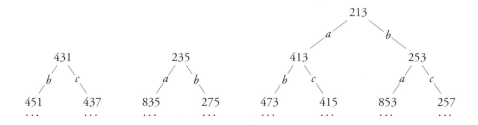

The third announcement is:

* *Cath: "I don't know my number."*

We now remove states where Cath would have known her number. These are $(2, 3, 5)$ and $(2, 1, 3)$. The state $(4, 3, 1)$ remains: It is indistinguishable for Cath from $(4, 3, 7)$.

The issue is whether we now have sufficient information to solve the problem. The fourth announcement was:

* *Anne: "I know my number. It is 50."*

We can determine from the tree—or rather, by now, from the forest of five trees—where Anne now would have known her number. The three triples where Anne now knows her number are

$$(4, 3, 1)$$
$$(4, 1, 3)$$
$$(8, 3, 5)$$

(It does not show from the picture, but before Cath's announcement the triple $(8, 3, 5)$ was indistinguishable for Anne from $(2, 3, 5)$.) Anne does not have 50 in any of $(4, 3, 1)$, $(4, 1, 3)$, and $(8, 3, 5)$. But as we mentioned before, the information analysis also holds for the tree wherein we multiply all numbers by a constant factor. So, the analysis is not merely for the tree with root $(2, 1, 1)$ but for any tree with root $(2x, x, x)$, where $x \geq 1$. The question now becomes whether 50 can be a multiple of 4 or 8, i.e., whether 4 or 8 are divisors of 50. They are not. We are stuck. Really?

6.3 The Solution

The tree with root $(2, 1, 1)$ is only one of the binary trees modeling the uncertainty of the three agents. And there are even more than the infinite number of trees with roots $(2x, x, x)$. There are two more types of binary tree, namely with root $(1, 2, 1)$ and with root $(1, 1, 2)$. And any of those with multiple arguments is also a binary tree in the model. Altogether, that describes the entire model. Because, take any triple (x, y, z) where one of the arguments is the sum of the other two, replace the largest number by the difference of the smaller two, and

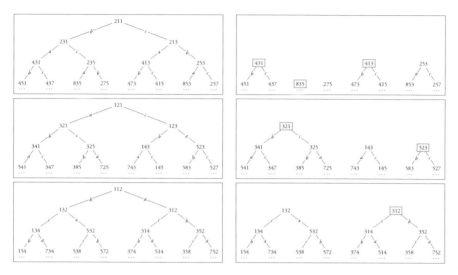

Fig. 6.1   The informative consequences of three ignorance announcements

repeat the procedure on the resulting triple. We will then always get a triple of shape $(2w, w, w)$, $(w, 2w, w)$, or $(w, w, 2w)$. (We apply what is known as the *Euclidean algorithm* to two of the three arguments x, y, z, where the w found is the greatest common divisor.) Turning the argument around: Any triple (x, y, z) is a multiple of a triple found in one of the trees with roots $(2, 1, 1)$, $(1, 2, 1)$, and $(1, 1, 2)$.

Unlike multiplying arguments in binary trees, swapping arguments makes a difference for the informative consequences of announcements. The other two trees are differently affected by the agents' announcements. In the tree with root $(1, 2, 1)$, only Bill knows what all numbers are. For any triple in that tree, it is already the case that Anne does not know her number. Therefore, if Anne says that she does not know her number, this does not result in a transformation of the binary tree; the announcement is not truly informative. Only the second announcement is informative. It then becomes known that the triple cannot be $(1, 2, 1)$, where Bill would have otherwise known his number. Cath's subsequent announcement is also informative. In the tree with root $(1, 1, 2)$, neither Anne's nor Bill's announcement is informative. Only Cath's announcement results in information change, namely the removal of triple $(1, 1, 2)$. Figure 6.1 reviews the three binary trees and the combined informative consequences of the three successive announcements. The states wherein Anne now knows her number are framed.

By now we are sufficiently informed to solve the problem. The triples wherein Anne knows her number are:

$$(4, 3, 1)$$
$$(8, 3, 5)$$
$$(4, 1, 3)$$
$$(3, 2, 1)$$
$$(5, 2, 3)$$
$$(3, 1, 2)$$

Given those six alternatives, $(5, 2, 3)$ is the unique triple such that Anne's number is a divisor of 50. Therefore, given that Anne now says,

*   *Anne: "I know my number. It is 50."*

Bill must have 20 and Cath must have 30. We are done! In fact, we are more than done, and we do not only know what the other two numbers are but also know who has which number. We conclude with some variations on this riddle.

6.4   Versions

A different formulation of the riddle is as follows:

Anne, Bill, and Cath all have a positive integer on their forehead. They can only see the foreheads of others. One of the numbers is the sum of the other two. All the previous is common knowledge. They now successively make the truthful announcements:

*   Anne: "I don't know my number."
*   Bill: "I don't know my number."
*   Cath: "I don't know my number."

What are the numbers, if Anne now knows her number, and if all numbers are primes?

Now, the answer should of course be: 5, 2, and 3. Or, more accurately, "Anne has 5, Bill 2, and Cath 3." Hans liked this version. He tested it on his friend Roy. Roy is smart. Without any doubt he immediately responded, "The numbers are 2, 3, and 5." Hans did not understand how Roy could have found this answer so quickly, given the nontrivial computations on trees needed to find it. Well, said Roy, "I simply gave you the first three primes, it seemed the most obvious answer." Right. One has to be careful about formulating riddles.

Puzzle 31 *You are being told the "Who has the sum" riddle by a friend, but for natural numbers* $(0, 1, 2, \ldots)$ *instead of positive integers* $(1, 2, 3, \ldots)$. *Show that it cannot be determined what the numbers are, if* 0 *is also allowed.*

Puzzle 32 *Consider the "Who has the sum" riddle for natural numbers (0 is also allowed). You are lazy and you want to find the solution by writing a computer program. In order to program it, the domain must be finite. You therefore, take an upper limit to the allowed number. Only triples* (x, y, z) *are allowed such that* $x, y, z \leq$ max, *where* max *is the upper limit, for example,* 10, *or* 21.

Having a maximum changes the nature of the riddle! For example, if the maximum is 10, *then if the numbers are* $(4, 9, 5)$. *Cath knows that she has* 5, *because the sum of* 4 *and* 9 *is* 13, *which is more than the maximum. Now if we set the maximum* too low, *then the three announcements can no longer be made truthfully. But if the maximum is too* high, *there will still be triples wherein Anne does not know the other numbers.*

For which maxima is it the case that after the three ignorance announcements by Anne, Bill, and Cath, Anne always knows her number?

6.5 History

The riddle appeared in the journal *Math Horizons,* in 2004, as "Problem 182" on page 324. This is a regular section of the journal with mathematical entertainment. See (Liu 2004). Hans van Ditmarsch first heard about the riddle in the version for natural numbers, when it cannot be solved: That version and the one with upper bounds (Puzzles 31 and 32) are his attempts to make sense of it prior to learning about the real puzzle. The bounds in Puzzle 32 were verified by the model checker DEMO (A Demo of Epistemic Modeling) with a script written by Ji Ruan (van Ditmarsch and Ruan 2007).

7

Sum and Product

A says to S and P: I have chosen two integers x, y with $1 < x < y$ *and* $x+y \leq 100$. *In a moment I will inform S of their sum* $s = x + y$, *and I will inform P of their product* $p = xy$. *These announcements will remain secret. You are required to make an effort to determine the numbers x and y.*

He does as announced. The following conversation now takes place:

1. *P says: I don't know the numbers.*
2. *S says: I knew you didn't know the numbers.*
3. *P says: Now I know the numbers.*
4. *S says: Now I also know the numbers.*

Determine x and y.

7.1 Introduction

The sum-and-product riddle can indeed be called a riddle, because the announcements made by S (for "sum") and P (for "product") do not seem very informative; they only talk about ignorance and knowledge and do not say anything about actual numbers. Still, these announcements are so informative that S and P can eliminate number pairs. For example, the numbers cannot be 2 and 3, or another prime number pair, because in all those cases P would immediately have derived the numbers from their product. In that case, he would not have been able to make the first announcement, "I don't know the numbers," truthfully. It is a bit harder to realize that the number can also not be, for example, 14 and 16. If that had been the case, then the sum would have been 30. This is also the sum of the prime numbers 7 and 23. If the product had been $7 \cdot 23$, then P would have known what the numbers are. In other words, if their sum is 30, S considers it possible that P knows what the numbers are. But S said, "I knew you didn't know the numbers." Therefore, the numbers cannot be 14 and 16.

By this kind of elimination of number pairs, S and P learn enough from their announcements to determine the unique solution of the problem. If you want to find this on your own, this is a good moment. But if you want more hints and explanations, first read the next section, and try again. (If you still have not found the solution, then read on all the way.)

7.2 I Know That You Do Not Know It

Puzzle 33 *A says to S and P: I have chosen two integers x, y with* $1 < x < y$ *and* $x + y \leq 10$. *In a moment I will inform S of their sum* $s = x + y$, *and I will inform P of their product* $p = xy$. *These announcements will remain secret. You are required to make an effort to determine the numbers x and y.*

He does as announced. The following conversation now takes place:

1. *P says: I don't know the numbers.*
2. *S says: Now I know the numbers.*
3. *P says: I still don't know the numbers.*

In the initial situation, the possible number pairs are: $(2, 3)$, $(2, 4)$, $(2, 5)$, $(2, 6)$, $(2, 7)$, $(2, 8)$, $(3, 4)$, $(3, 5)$, $(3, 6)$, $(3, 7)$, $(4, 5)$, and $(4, 6)$. Some of those have the same sum, such as $(2, 5)$ and $(3, 4)$. Exactly two of those have the same product, namely $(3, 4)$ and $(2, 6)$. We can depict this in a grid, as below. Number pairs with the same sum are connected by a solid line (a sum line). Number pairs with the same product are connected with a dashed line (a product line).

$(2, 8)$

$(2, 7)$ $(3, 7)$

$(2, 6)$ $(3, 6)$ $(4, 6)$

$(2, 5)$ $(3, 5)$ $(4, 5)$

$(2, 4)$ $(3, 4)$

$(2, 3)$

If P now says, "I don't know the numbers," then we eliminate all number pairs wherein P would have known the numbers. For example, if the number pair is $(2, 3)$, then their product is 6, and P knows that the numbers are 2 and 3. But he said he did not know. Therefore, $(2, 3)$ cannot be the number pair. However, if the number pair is $(3, 4)$, then P cannot distinguish it from pair $(2, 6)$. In all other cases (such as $(2, 3)$), P knows what the numbers are. The result of processing P's announcement is therefore:

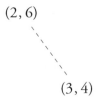

$(2, 6)$

$(3, 4)$

The announcement that P made is informative for S. If the numbers are 2 and 6, then before P's announcement S could not distinguish pair $(2, 6)$ from pair $(3, 5)$, as both have sum 8, but after P's announcement, S now knows that the numbers are 2 and 6. Similarly, if the numbers are 3 and 4, then before P's announcement S could not distinguish pair $(3, 4)$ from pair $(2, 5)$, because both add up to 7, but after P's announcement S now knows that the numbers are 3 and 4.

Following P's announcement, S says, "Now I know the numbers." That is not informative for P: He already knew that. But P can still not distinguish $(2, 6)$ from $(3, 4)$. Therefore, P says, "I still don't know the numbers . . . ."

7.3   I Knew You Did Not Know

In the original riddle, S says, "I knew you didn't know," i.e., S knew that P did not know the numbers. The past tense indicates that the announcement applies to the initial state of information, before processing P's announcement "I don't know the numbers." In the initial state of information, it should therefore hold that "S knows that P does not know what the numbers are."

This statement is false for all number pairs of the model wherein the sum is at most 10. For example, if the sum is 8, then S cannot distinguish $(2, 6)$ from $(3, 5)$. So, if the pair is $(2, 6)$, P does *not* know what the numbers are, whereas if the pair is $(3, 5)$, P *does* know what the numbers are. Therefore, S considers both possible and does not know whether P knows the numbers.

Puzzle 34 *Take the original version of the riddle. Show that S knows that P does not know the numbers if the sum is* 11.

We have to determine for all number pairs with sum 11 whether there is another number pair with the same product. The possible number pairs (also showing the alternatives with the same product) are:

Pair with sum 11	Product of this pair	Other pair with the same product
(2,9)	18	(3,6)
(3,8)	24	(4,6), (2,12)
(4,7)	28	(2,14)
(5,6)	30	(2,15), (3,10)

Now suppose the number pair is $(2, 9)$. Then S cannot distinguish pair $(2, 9)$ from the three other pairs with sum 11. In those cases, P also does not know what the numbers are, because there is then another pair with the same product. Therefore, if the number pair is $(2, 9)$, it is true that "S knows that P does not know what the numbers are."

Visually, a solid line (*sum line*) connecting all pairs with sum 11 should in any node intersect with a dashed line (*product line*) to another pair with the same product as that node. To make it clear, we scale the sum lines and product lines in a grid, in Figure 7.1. On the left, we show the entire grid. In the middle, we only show the sum line 11 and the four intersecting product lines. On the right, we show a visual simplification of that.

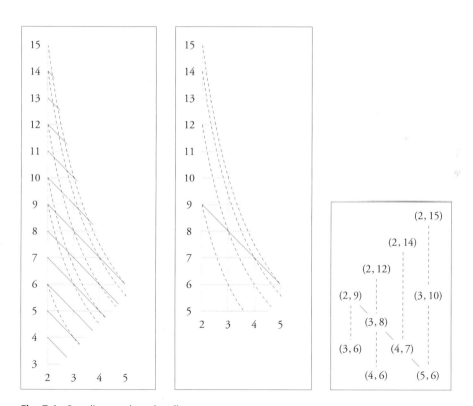

Fig. 7.1 Sum lines and product lines

7.4   Solution of Sum and Product

The four announcements are:

1. *P says: I don't know the numbers.*
2. *S says: I knew you didn't know the numbers.*
3. *P says: Now I know the numbers.*
4. *S says: Now I also know the numbers.*

The First Announcement

* *P says: I don't know the numbers.*

It is superfluous in the analysis, because the second announcement employs the past tense, "knew." This entails that the second announcement needs to be processed in the initial state of information, and in that case it also implies the first announcement. We now subsequently analyze the three pertinent announcements.

The Second Announcement

* *S says: I knew you didn't know the numbers.*

There are ten sums of two numbers such that S knows that P does not know the numbers. One of those is 11, as we have already demonstrated in the previous subsection. To determine systematically, if S knows this, we have to do this for all possible sums. Because $1 < x < y$ and $x + y \leq 100$, that means *all* sums between 5 and 100! That seems a lot. But it is not as bad as it seems.

In the first place, we can rule out all even sums except $(2, 4)$. This is because of the famous "Goldbach conjecture" that every even number is the sum of two prime numbers. For the numbers smaller than 100, this is not a conjecture but true. In all such cases, S therefore considers it possible that P knows the numbers (because their product would then be the unique product of those two prime numbers). *Almost* all even sums can be eliminated: If the sum is 6, then S and P both know that the number pair is $(2, 4)$. The number 6 is the sum of two identical prime numbers, namely $3 + 3$. But x and y in $x + y$ have to be different. However, all even numbers larger than 6 are the sum of more than one prime number pair. So, one of those pairs must consist of different prime numbers.

There are more tricks like that:

We can eliminate all odd sums $q + 2$ such that q is a prime number (it is more common to call prime numbers p instead of q, but p already stands for the product of x and y), because for pair $(2, q)$ there is then a unique product $2q$.

If the sum $x + y$ is larger than a prime number q over 50, then $(x + y - q, q)$ gives a unique product, because if q would not be one of two numbers with product $(x + y - q)q$, then one of the numbers must be q multiplied by a prime factor of $x + y - q$, and therefore it must be at least $2q$, which would be more than 100. This cannot be, as the sum of an alternative number pair with the same product must also be at most 100. That eliminates all sums over 55, as 53 is a prime number.

The following sums satisfy S's announcement:

$$11, 17, 23, 27, 29, 35, 37, 41, 47, 53$$

For sum 11, we refer to Puzzle 34 above. Next comes sum 17. We list all number pairs with sum 17, and for each such pair the other pairs with the same product. For the other sums, we leave the verification to you.

Pair with sum 17	Product of this pair	Other pair with same product
(2,15)	30	(3,10), (5,6)
(3,14)	42	(2,21), (6,7)
(4,13)	52	(2,26)
(5,12)	60	(2,30), (3,20), (4,15), (6,10)
(6,11)	66	(2,33), (3,22)
(7,10)	70	(2,35), (5,14)
(8,9)	72	(2,36), (3,24), (4,18)

We continue by schematically representing the informative consequences of the second announcement in a figure. First, we need to model the initial state of information. In the picture, we only show some number pairs, and we also do not depict it on the proper scale, all this only to improve readability. Pairs with equal sums are found on the diagonal in the diagram. They are connected with solid lines, and just as in the previous section, we call them *sum lines*. For each of the 10 remaining sums, there is such a sum line, but in the picture we only show the sum lines for sums 11, 13, 17, and 23. Equal product pairs are connected with dashed lines that we call *product lines*. Their pattern is less predictable. (The intersections of sum lines and product lines are nothing but the integer intersection points—we have an $\mathbb{N} \times \mathbb{N}$ grid—of the hyperbolas $xy = p$ and the diagonals $x + y = s$.)

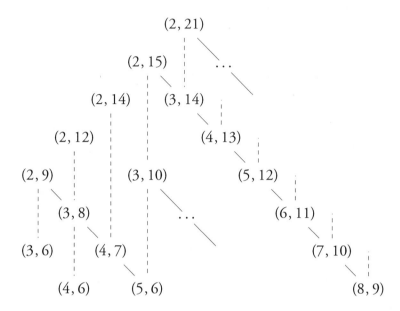

When S now says, "I knew you didn't know," all sum lines (i.e., all number pairs constituting those sum lines) are eliminated except those for the 10 sums 11, 17, 23, etc. For example, above, the line with sum 13 is now removed (as $3 + 10$ is also $2 + 11$, the sum of two primes). We also remove all product lines that intersect with none or only one of the remaining sum lines. (If it does not intersect anyway, fine: All eliminated pairs are on removed sum lines. If it intersects with only one of the ten remaining sum lines, a pair with that product indeed remains on that sum line but no product *line,* as for a line we need at least two nodes. . .) The resulting model, schematically again, is as follows. For example, the removed product lines include the one containing $(4, 13)$, for which the only alternative pair with the same product is $(2, 26)$, also the sum of the prime numbers 5 and 23.

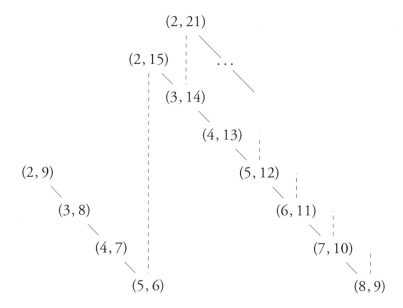

The Third Announcement

> * *P says: Now I know the numbers.*

In the model wherein this announcement is made, there are two kinds of number pairs: Those such that their product is *not* the same as the product of a number pair on another sum line, and those such that their product is the same as the product of a number pair on another sum line. Let us call the the pairs without such an alternative *closed,* and the pairs that have such an alternative equal product *open.*

For the closed number pairs, *P* knows what the numbers are, for the open number pairs *P* does not know what the numbers are. Because *P* says that he now knows the numbers, all open number pairs will be eliminated by this announcement.

The line with sum 11 has one open pair, (5, 6), that is eliminated by the third announcement, and the other three are closed, so they remain.

For the line with sum 17 (see also the previous table), the following alternatives were available in the information state where the third announcement was made:

Pair with sum 17	Product of this pair	Other pair with same product
(2,15)	30	(5,6)
(3,14)	42	(2,21)
(4,13)	52	-
(5,12)	60	(3,20)
(6,11)	66	(2,33)
(7,10)	70	(2,35)
(8,9)	72	(3,24)

From sum line 17, therefore only the pair $(4, 13)$ remains after the third announcement.

We now did 11 and 17. The same process of elimination of open pairs can be applied to the remaining eight sum lines. Schematically, the result is as follows:

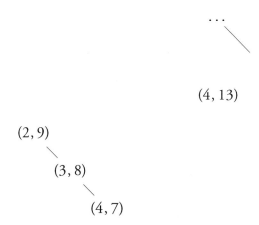

The Fourth Announcement

* *S says: Now I also know the numbers.*

After the third announcement, there is one sum line from which only one number pair remains: This is the sum 17 and the pair $(4, 13)$. All other sum lines contain more than one pair. We have seen that for sum 11 the pairs $(2, 9)$, $(3, 8)$, and $(4, 7)$ remain. For yet another example, for sum line 23 these are (at least) $(4, 19)$ and $(10, 13)$. The verification of the remainder we leave to the reader again. All sum lines containing more than one pair are eliminated by the fourth announcement. The numbers should therefore be 4 and 13! We solved the riddle.

It is most remarkable that from a very large number of possible number pairs only one remains by a process of announcements that only indirectly refer to these numbers, by way of knowledge and ignorance statements. Merely three announcements are sufficient to reduce thousands of possible number pairs to just one pair, the unique solution.

7.5   Versions

Puzzle 35 *In this version, the third announcement, by P, is not that he knows the numbers, but that he does not know the numbers. After that, S knows. And after that, P can then determine the numbers.*

   We recall that in the original riddle the first announcement was superfluous for you to determine the numbers (it was subsumed by the second announcement). The last announcement by P in this version of the riddle is also not necessary for you to determine the numbers (different from the last announcement, by S, in the original version of the riddle). You can already determine what the numbers are after the fourth announcement.

   A says to S and P: I have chosen two integers x, y with $1 < x < y$ and $x+y \leq 100$. In a moment I will inform S of their sum $s = x + y$, and I will inform P of their product $p = xy$. These announcements will remain secret. You are required to make an effort to determine the numbers x and y.

   He does as announced. The following conversation now takes place:

1. *P says: I don't know the numbers. (superfluous)*
2. *S says: I knew you didn't know the numbers.*
3. *P says: I still don't know the numbers.*
4. *S says: Now I know the numbers.*
5. *P says: Now I know the numbers. (superfluous)*

   Determine x and y.

Puzzle 36 *Suppose that in Puzzle 33, wherein the sum of the numbers is at most 10, it is also permitted that the numbers are equal: $x = y$. Show how the model and the subsequent information change are affected by this.*

Puzzle 37 *Suppose that in the original version of the riddle it is allowed that $x = y$. This changes the uncertainty of S and P about the numbers in the initial model, and also, in principle, during the subsequent information processing. However, show that after the second announcement the resulting model already is the same (and therefore, also after the remaining two announcements).*

7.6   History

The well-known topologist Hans Freudenthal published the sum-and-product riddle in "Nieuw Archief voor Wiskunde" (New Archive of Mathematics), a Dutch-language mathematical journal, in 1969. It is riddle no. 223 in the last issue of "Nieuw Archief voor Wiskunde" in 1969 (Freudenthal 1969, p. 152).

> No. 223.   *A* zegt tot *S* en *P*: Ik heb twee gehele getallen x, y gekozen met $1 < x < y$ en $x + y \leq 100$. Straks deel ik $s = x + y$ aan *S* alleen mee, en $p = xy$ aan *P* alleen. Deze mededelingen blijven geheim. Maar jullie moeten je inspannen om het paar (x, y) uit te rekenen.
>
> Hij doet zoals aangekondigd. Nu volgt dit gesprek:
>
> 1.   *P* zegt: Ik weet het niet.
> 2.   *S* zegt: Dat wist ik al.
> 3.   *P* zegt: Nu weet ik het.
> 4.   *S* zegt: Nu weet ik het ook.
>
> Bepaal het paar (x, y).   *(H. Freudenthal)*

This is the English translation:

> No. 223.   *A* says to *S* and *P*: I have chosen two integers x, y with $1 < x < y$ and $x + y \leq 100$. In a moment I will inform only *S* of their sum $s = x + y$, and I will inform only *P* of their product $p = xy$. These announcements remain secret. You are required to make an effort to determine the pair (x, y).
>
> He does as announced. The following conversation now takes place:
>
> 1.   *P* says: I don't know it.
> 2.   *S* says: I knew you didn't know it.
> 3.   *P* says: Now I know it.
> 4.   *S* says: Now I also know it.
>
> Determine the pair (x, y).   *(H. Freudenthal)*

We have stayed very close to the original version in this chapter.

In the subsequent issue of "Nieuw Archief," in 1970, various solutions were discussed (Freudenthal 1970). Those who solved the problem were named. Interestingly, many names later became prominent in the mathematics and computer science community in the Netherlands. The sum-and-product riddle then resurfaced in other parts of the world. McCarthy (1990), one of the founding fathers of artificial intelligence, wrote in the late 1970s on the sum-and-product riddle. This was only formally published later. In the same work, he treats the muddy children problem. His formulation is as follows:

Two numbers *m* and *n* are chosen such that $2 \leq m \leq n \leq 99$. Mr. *S* is told their sum and Mr. *P* is told their product. The following dialogue ensues:

1. Mr. *P*: I don't know the numbers.
2. Mr. *S*: I knew you didn't know. I don't know either.
3. Mr. *P*: Now I know the numbers.
4. Mr. *S*: Now I know them too.

In view of the above dialogue, what are the numbers?

The Freudenthal original and the McCarthy version are somewhat different:

* In McCarthy, *both numbers* are at most 99, whereas in Freudenthal *their sum* is at most 100.
* In McCarthy, the numbers may be *the same,* in Freudenthal they may *not*.
* In McCarthy, the second annoucement, by Mr. *S, also contains "I don't know either,"* but this extra information is *not* in Freudenthal.

The McCarthy version therefore allows many more number pairs in the initial state, and has a second announcement that is, in principle, more informative than in the Freudenthal version. None of these changes, neither in isolation nor combined, make any difference for the solution of the riddle. (All this has been checked, and checked again, in the very extensive literature on the riddle. See also Puzzle 37.)

From the 1970s onward, many different versions of Freudenthal's sum-and-product riddle have been proposed. The variations involve other announcements and other ranges of numbers. The version of Puzzle 35 is our own. For the many versions of the sum-and-product riddle in the literature on recreational mathematics, see, for example, Gardner (1979), Sallows (1995), Isaacs (1995), and Born et al. (2006, 2007, 2008), and the website www.mathematik.uni-bielefeld.de/~sillke/PUZZLES/logic_sum_product. We recommend the succinct and elegant analysis by Isaacs (1995).

It is not known how the riddle got from Freudenthal to McCarthy—McCarthy did not know about the Freudenthal origin when he wrote about the riddle, and Gardner also did not know this (personal communication). McCarthy had found it on the bulletin board of Xerox PARC, an institute near Stanford University where he worked, some 10 years after the Freudenthal version was published. The search for the missing link from Freudenthal to McCarthy has been reported (in Dutch) by van Ditmarsch et al. (2009). The riddle was prominently present in the first publication on so-called public announcement logic by Plaza (1989), and another review of different versions of the riddle, including such epistemic logical analyses, is by van Ditmarsch et al. (2007).

8

Two Envelopes

A rich man gives you two indistinguishable envelopes, each of which contains a positive sum of money. One envelope contains twice as much money as the other envelope. You may choose one envelope and keep the money. You pick one envelope, open it, and it contains 100 dollars. Now the rich man offers you the option to take the other envelope instead. What should you do?

8.1 High Expectations

What should you base your decision on? Mathematicians developed a framework for this, called probability theory. You can use the probabilities of different outcomes to compute the *expected value* of a given decision. Suppose someone throws a coin, if it is heads, you get 6 dollars and if it is tails, you pay 4 dollars. On the assumption that the odds between heads and tails are equal (50 % of the time it comes up heads and 50 % of the time it comes up tails), you can figure out that the expected value of a throw of the coin is 1 dollar. In other words: if you play the game often enough, your average gain per game will be 1 dollar. Suppose you play it 100 times, and you win exactly 50 of those and lose exactly 50 of those. Then you win 300 dollars, but you also pay 200 dollars. So altogether you win 100 dollars, 1 dollar per game. Of course you might instead have won 47 games and lost 53 games, and you would still have won something, but even more unequal outcomes are increasingly unlikely, for example, it is very improbable that you win 30 and lose 70 if the coin is fair.

Hence, to decide whether you should switch envelopes, you can use probability theory. Probability theory tells you to take the envelope with the highest expected value. Obviously, the expected value of the envelope in your hands is its actual value, 100 dollars. What is the expected value of the other envelope? It can contain one of two possible amounts. It will contain either 50 dollars or 200 dollars and both seem equally probable. So, you have a probability of 50% to win 50 dollars and a probability of 50% to win 200 dollars. The expected value is therefore 125 dollars: $\frac{1}{2} \cdot 50 + \frac{1}{2} \cdot 200 = 125$. As this is more than the

100 dollars in your hands, in the current envelope, you should therefore take the other envelope.

Although probability theory appears to advise you to switch envelopes, this seems to contradict common sense. Because no matter what amount we find in the envelope now in our hands, the expected value of the other envelope will always be higher. Suppose our initially chosen envelope contains x dollars. Then the other envelope contains $\frac{1}{2}x$ or $2x$ dollars. So the expected value of the other envelope is $1\frac{1}{4}x$ (we have that $\frac{1}{2} \cdot \frac{1}{2}x + \frac{1}{2} \cdot 2x = 1\frac{1}{4}x$). If the other envelope always has a higher expected value, then you might as well choose the other envelope initially. No need to change. But then, from the perspective of that envelope, the original initially chosen envelope would now always have a higher expected value! It appears that, no matter what you choose, you can never be satisfied with your first choice.

8.2   A Subtle Error

The argument compelling you to switch envelopes contains a subtle error. The assumption is incorrect that the probability for the other envelope to contain 50 dollars is equal to the probability for the other envelope to contain 200 dollars. This does not have to be so. Assume that the rich man is a bit of a scrooge and always fills the envelopes with a 50 dollar bill and a 100 dollar bill, whenever he plays this game. Then the other envelope will always contain 50 dollars. Whereas, if he is more generous and always fills the envelopes with 100 dollars and 200 dollars, then there will definitely be 200 dollars in the other envelope. The rich man's policy in filling envelopes determines the expected outcome of the other envelope when you find 100 dollars in 1 envelope.

Consider the following distribution policy: the rich man will throw a coin. If it is heads, then he fills the envelopes with 50 and 100 dollars. If it is tails, he fills the envelopes with 100 and 200 dollars. Now, if you draw the envelope with 100 dollars, it is indeed the case that the probability of the other envelope containing 50 is equal to the probability of the other envelope containing 200. So now, you should change envelopes. But, of course, it is now no longer the case that if you had chosen the other envelope, you would also have been advised to change. If you had found 200 in that envelope, you would not have changed, whereas if you had found 50 in that envelope, you would have changed. Either way, you would have known that the other envelope contained 100 dollars.

So the justification for switching envelopes is wrong. Still, you have to make a choice. Should you change or not? Unfortunately, we cannot answer this question. As long as we do not know how the rich man came to fill the two envelopes with money, we cannot say anything about this. If we had known this, we could have said something sensible about it, as above, but not without any information at all.

8.3 Versions

This version of the two-envelope problem is called the Ali-Baba version. It is by Nalebuff (1989).

Puzzle 38 *A rich man promises you the contents of one of two indistinguishable envelopes. He fills one envelope with money, and he gives it to you. Then, he throws a (one of a pair of) dice, in secret. If the outcome is odd, he fills the other envelope with half that amount of money; if it is even, with double the amount. The rich man now asks you if you wish to change envelopes. What should you do?*

8.4 History

The puzzle is also called a paradox, because it is such a complex problem. Lots of people, including mathematicians, are indeed puzzled by it.

The earliest known version of the paradox is on ties and is in a booklet by Kraitchik (1943).

Each of two persons claims to have the finer necktie. They call in a third person who must make a decision. The winner must give his necktie to the loser as consolation. Each of the contestants reasons as follows: "I know what my tie is worth. I may lose it, but I may also win a better one, so the game is to my advantage." How can the game be to the advantage of both?

Kraitchik also discusses a version wherein two people compare the number of coins in their wallet. This form is also found in Gardner (1982). It is unclear how the form presented in this chapter came about. Zabell (1988a, b) heard it from Nalebuff (1989), but this is as far as the history of the riddle goes.

The solution presented in this chapter is based on Albers et al. (2005).

9

One Hundred Prisoners and a Light Bulb

A group of 100 prisoners, all together in the prison dining area, are told that they will be all put in isolation cells and then will be interrogated one by one in a room containing a light with an on/off switch. The prisoners may communicate with one another by toggling the light switch (and that is the only way in which they can communicate). The light is initially switched off. There is no fixed order of interrogation, or interval between interrogations, and the same prisoner may be interrogated again at any stage. When interrogated, a prisoner can either do nothing or toggle the light switch, or announce that all the prisoners have been interrogated. If that announcement is true, the prisoners will (all) be set free, but if it is false, they will be executed. While still in the dining room, and before the prisoners go to their isolation cells (forever), can the prisoners agree on a protocol that will set them free?

9.1 How to Count to a Hundred with Only 1 Bit?

The riddle seems unsolvable. Only 1 bit is available for information transmission: the light can be on or off. But there are 100 prisoners. The number 100 is between 64 and 128. Its binary representation, therefore, takes 7 bits. And we are not even talking about the protocol to solve the riddle! How can 1 bit be sufficient to do all of this?

A common procedure in mathematics is to generalize a solution for small numbers to one for larger numbers. This is because it looks like a so-called proof by natural induction: if we can show that something holds for a basic case (like one prisoner, or two prisoners), and if, given a proof for the case n (such as n prisoners), we can prove it for the case $n + 1$, then it holds for all natural numbers n (from the basic case onward). In the prisoners riddle, natural induction seems a stumbling block for the intuition. We can easily find a protocol for one prisoner, and for two prisoners, but the step from two to three prisoners seems unbridgeable. Let us take you there step by step, prisoner by prisoner: we solve it for one prisoner, then for two, then we try it for more than two.

9.2   One Prisoner

Let there be one prisoner: Anne. The first time Anne is interrogated, she announces that everybody has been interrogated. She does not need the light bulb for that. Our first attempt is, therefore, as follows:

Protocol 1 *If you are being interrogated, you announce that everybody has been interrogated.*

9.3   Two Prisoners

Protocol 1 does not work if there are more than one prisoner. But, maybe, we can adapt it to the case of two prisoners. Suppose there are two prisoners: Anne and Bob. The first to be interrogated turns on the light. Without loss of generality we can assume that this is Anne. The next person to be interrogated can be either Anne again or Bob. If it is Bob, then he sees that the light is on and, therefore, knows that Anne must already have been interrogated (only the prisoners themselves may turn the light on or off). So, Bob can then truthfully announce that everybody has been interrogated. Anne and Bob go free. If the next person to be interrogated is Anne, then, as the light is on, she may then be tempted to think that Bob has not been interrogated yet, because in that case surely he would have announced that they would both have been interrogated and she would already have been free. We are a bit vague: Anne's reasoning is based on what she finds probable that Bob would have done. But there is nothing against explicitly agreeing on such behavior prior to the interrogation, i.e., agreeing on a protocol. We get this:

Protocol 2 *If you are being interrogated and the light is off, turn it on; if you are being interrogated and the light is on and you have turned it on before (i.e., during a prior interrogation), do nothing; if you are being interrogated and the light is on but you have not turned it on before, announce that everybody has been interrogated.*

9.4   A Protocol for Three Prisoners?

Now suppose that there are three prisoners: Anne, Bob, and Caroline. This is harder. Anne, again the first to be interrogated, may turn on the light again. Now suppose Bob is interrogated. He could turn it off again. If, after that,

Anne is interrogated, she would then know, as before, that at least two prisoners have been interrogated. We are making progress! Now suppose Caroline is interrogated next. Unfortunately, Caroline cannot conclude now that another prisoner has already been interrogated. As the time between interrogations is unknown, Caroline considers it also possible that she is the first to be interrogated, in which case the light would also have been off. What should she do? Well, that really is nothing else but asking yourself what Anne should have done at the first interrogation. Let us continue for a while in this fantasy (it is fantasizing indeed, because this is going to be a dead end). The light only represents 1 bit, but the prisoners themselves can count. They can count how often they turn on the light and how often they turn off the light. Maybe we can use this in a protocol? Consider this:

Protocol 3 (Incorrect!) *The first time that you are interrogated and that you observe that the light is off, turn it on. If you are interrogated and the light is off and you have turned it on before, do nothing. If you are interrogated and the light is on and you know that you have not turned it on, then turn it off. The second time that you turn the light off, you announce that all prisoners have been interrogated.*

This protocol sometimes gets us what we want, and sometimes not. First consider the interrogation sequence

$$\text{Anne} - \text{Bob} - \text{Bob} - \text{Anne} - \text{Caroline} - \text{Anne}.$$

Let 0 mean that the light is off and 1 that the light is on. We can represent the evolution of the system by recording the state of the light as an upper index, so that we get

$$^0\text{Anne}^1\text{Bob}^0\text{Bob}^1\text{Anne}^0\text{Caroline}^1\text{Anne}^0$$

and when we also record how often a prisoner turns it off, by a lower index, we get

$$^0\text{Anne}_0^1\text{Bob}_1^0\text{Bob}_1^1\text{Anne}_1^0\text{Caroline}_0^1\text{Anne}_2^0.$$

Anne correctly announced that all the three prisoners have been interrogated. Here, we get what we want.

Now consider the interrogation sequence starting with

$$\text{Anne} - \text{Bob} - \text{Bob} - \text{Caroline} - \text{Caroline} - \text{Anne} -$$

Using the same annotations we get

$$^0\text{Anne}_0^1\text{Bob}_1^0\text{Bob}_1^1\text{Caroline}_1^0\text{Caroline}_1^1\text{Anne}_1^0 \ldots$$

At this stage all the three prisoners have turned off the light once; also, they have turned on the light once. But they only turn it on once. So no matter who is going to be interviewed next and no matter how often, the light will remain off forever. Here, we do not get what we want.

Well, what if you turn it on twice instead of once, just as you turn it off twice? Then consider the sequence

$$^0\text{Anne}_0^1\text{Bob}_1^0\text{Bob}_1^1\text{Anne}_1^0\text{Bob}_0^1\text{Anne}_2^0.$$

Anne now announces incorrectly that all the three have been interrogated and they will get hanged. Further variations on this theme lead into similar trouble.

Now in the sequence where the announcement will never be made, like the one above, you could say that by the time a prisoner has been interrogated even ten more times (and everybody can count!), it becomes very likely that everybody else has already been interrogated. So a prisoner might make an *informed guess* that everybody has been interrogated and announce it. It becomes reasonable to *believe* that everybody has been interrogated at some stage. But belief is not knowledge. The prisoner might be mistaken. In order to *know*, without a shadow of doubt, that everybody has been interrogated, we need another kind of protocol.

9.5 No Tricks

You may have been considering various sorts of tricks in the mean time. Well, the riddle is not really a tricky question. Let us rule out some tricks.

* Suppose the light is off but it is still warm, then somebody must have just turned it off! *It is not allowed to touch the light bulb to find out whether it is cold or warm.* And what difference would it make anyway? How would that get you to count to 100 prisoners?
* Let us smash the light if you see that it is on for the second time, so that a prisoner seeing the smashed light, and who has not turned off a light yet, can announce that everybody has been interrogated. Well, this would solve it for three prisoners indeed. Now try four prisoners! *It is not allowed to smash the light.* Anyway, if you smash the light, even the solutions we know about can no longer be executed, so, really, better not.
* Keep track of the time! Sorry: As a Kiwi would say "Been there, done that." *The interval between interrogations is variable.* The first prisoner may be interrogated 10 times consecutively in the first hour. Or only for the first time after 3 days, and that may then still have been the first interrogation overall. *There is no point whatsoever for any prisoner to keep the track of the time.*

* Maybe the prisoners cannot see who is being interrogated but they still can see whether the light is on or off from their isolation cells. *The interrogation room cannot be seen from the isolation cells housing the prisoners.* Now guess what: therefore they are called "isolation cells" in the riddle! Surely, that cannot be a big surprise.
* *There is no secret connection between the light switch and any of the isolation cells where the other prisoners are.*
* *From an isolation cell, you cannot hear that in the interrogation room the light switch is turned on or off.* Really, do we have to explain "isolation" again?

All the prisoners may have different names, or be identified with consecutive numbers from 1 to 100, or whatever. It all does not matter.

9.6   Solution for One Hundred Prisoners

The solution comes closer once you realize that not every prisoner needs to have the same *role* in the protocol. The prisoners can agree on a protocol as long as they are still in the prison dining room, prior to being separated and going to their isolation cells. They can be given different tasks in a protocol. Counting how often you turn off the light is quite useful as long as all the prisoners know that you are the only prisoner making that count. This is embodied in the following protocol.

Protocol 4 *The prisoners appoint one amongst them as the counter. All non-counting prisoners follow this protocol: the first time they enter the room when the light is off, they turn it on; on all other occasions, they do nothing. The counter follows a different protocol: if the light is off when he enters the interrogation room, he does nothing; if the light is on when he enters the interrogation room, he turns it off; when he turns off the light for the 99th time, he announces that everybody has been interrogated.*

It will be clear to the reader that this protocol indeed works. These are the three executions for the case of three prisoners, where Anne is the counter. Again, the upper index represents the state of the light, and the lower index represents how often Anne has turned it off. (The other prisoners do not need to keep count.)

1. $^0\text{Bob}^1\text{Anne}_1^0\text{Caroline}^1\text{Anne}_2^0$
2. $^0\text{Anne}^0\text{Bob}^1\text{Caroline}^1\text{Anne}_1^0\text{Bob}^0\text{Anne}_1^0\text{Caroline}^1\text{Caroline}^1\text{Bob}^1\text{Bob}^1\text{Anne}_2^0$
3. $^0\text{Bob}^1\text{Anne}_1^0\text{Bob}^0\text{Caroline}^1\text{Bob}^1\text{Anne}_2^0$

Suppose that Anne is the counter and that from the 100 prisoners only Anne and Bob will be interrogated, alternatingly. Then it will never happen that Anne can say that everybody has been interrogated. Does the protocol therefore not work? It is correct that in that scheduling of interrogations the solution will never come closer. The final announcement will never been made. But that is because not all prisoners have been interrogated! A condition for termination of Protocol 4 is so-called "fairness" or "liveness" of the scheduling of interrogations, that was formulated in the riddle as

(...), and the same prisoner may be interrogated again at any stage.

where by "may be" we mean "will be with nonzero probability." If that condition holds, then the counter will truthfully announce at some future interrogation that everybody has been interrogated.

There are many schedulings where all prisoners are interrogated but that are not fair. We can imagine that in the first 100 interrogations everybody gets interrogated in turn, and after that only Anne and Bob, alternatingly:

$$0, 1, 2, \ldots, 98, 99, 0, 1, 0, 1, \ldots$$

This scheduling is unfair. The condition "the same prisoner may be interrogated again at any stage" has not been fulfilled. After the first 100 interrogations ("at any stage") only Anne and Bob will be interrogated.

When it holds that "the same prisoner may be interrogated again at any stage," then every prisoner will be interrogated infinitely often. This is why: Take any moment during the interrogation sequence. After that, Anne will be interrogated a next time. Take that moment. After that, Anne will be interrogated again. And so on.

9.7 Versions

The State of the Light Suppose it is not known whether the light is initially on or off... Protocol 4 does not work now, not even if we let the counter count one more. For example, for the three prisoners, consider the following four executions. When Anne makes the announcement her name is in bold font.

1. ^{1}Anne$^0_1$Caroline^1**Anne**$^0_2$ Anne counts to 2. **Wrong**
2. ^{1}Anne$^0_1$Caroline^{1}Anne$^0_2$Bob^1**Anne**$^0_3$ Anne counts to 3. **Right**
3. ^{0}Bob^{1}Anne$^0_1$Caroline^{1}Anne$^0_2$Bob^0**Anne**$^0_2$ Anne counts to 2. **Right**
4. ^{0}Bob^{1}Anne$^0_1$Caroline^{1}Anne$^0_2$Bob^{0}Anne$^0_2 \ldots$ Anne counts to 3. **Wrong**

According to Protocol 4 Anne should count until 2. Execution 3 is an example where her announcement is correct. But in Execution 1 she incorrectly announces that everybody has been interrogated, when she is interrogated for the second time. So that does not work. Now let Anne count to 3 instead. In Execution 2 she is now able to make a correct announcement. But in Execution 4 she will never make an announcement, no matter how often Anne, Bob, and Caroline are subsequently interrogated: Bob and Caroline have both turned on the light already, and Anne turned off the light for both of them, and it will remain off forever now. What to do?

Puzzle 39 *Suppose it is not known whether the light is initially on or off. Give a protocol to solve the problem.*

Using Knowledge of the Non-counting Prisoners   Let us look again at the three different executions of Protocol 4 for the three prisoners, above. In Execution 1, everybody has been interrogated after three interrogations, and Anne knows that at the fourth interrogation. In Execution 2, again, everybody has been interrogated after three interrogations, but it takes another eight interrogations before counter Anne knows that and makes the announcement. Can it be done faster? Yes, it can. We list the three executions once more, with in boldface the first agent to know that all have been interrogated, at the moment he or she learns that.

1. ^{0}Bob^{1}Anne$_1^0$Caroline^1**Anne**$_2^0$
2. ^{0}Anne^{0}Bob^{1}Caroline^{1}Anne$_1^0$Bob^{0}Anne$_1^0$**Caroline**^{1}Caroline^{1}Bob^{1}Bob^1
     Anne$_2^0$
3. ^{0}Bob^{1}Anne$_1^0$Bob^{0}Caroline^1**Bob**^{1}Anne$_2^0$

This is why they get to know it first:

1.   In Execution 1, Anne is indeed the first to know that everybody has been interrogated.
2.   In Execution 2, Caroline learns that everybody has been interrogated at her second interrogation, before Anne knows that. In her first interrogation she notices that the light is on: Bob must have done that. In her second interrogation she sees that the light is off: Anne must have done that. Therefore, everybody has been interrogated.
3.   In Execution 3, Bob knows that everybody has been interrogated before Anne knows that. The first time he is being interrogated, he turns on the light. In his second interrogation, the light is off, so he does nothing, as he has turned it on before. But, because he observes that the light is off, he knows that Anne must have been interrogated and that she has turned off

the light. In his third interrogation he sees that the light is on, again. He leaves it on, according to protocol. But he knows that only Caroline can have turned on the light. Therefore, he can now announce truthfully that everybody has been interrogated.

This is embodied in the following protocol. We give it in a version for n prisoners, where $n \geq 3$. As all prisoners now count, it is a bit of a misnomer to continue to call the second role that of the "non-counter." Never mind.

Protocol 5 *The counter does as in Protocol 4. The non-counters also do all they do in Protocol 4, but they also do more: they count how often they observe (*) a change of the light from off to on. When they observed this $n - 1$ times, they announce that all prisoners have been interrogated.*

Now in order for a succinct presentation of this protocol, we were vague about what "observe a change of the light from off to on" means: (*) when a non-counting prisoner finds the light on during his first interrogation, this counts as observing a change from off to on; when a non-counting prisoner turns on the light, this counts as observing a change from off to on, and otherwise, when a non-counting prisoner observes the light off during one interrogation (without turning it on) and observes the light on during a later interrogation, this also counts as observing a change from off to on.

You may now check that Execution 2 and Execution 3 above indeed terminate according to this protocol, at the second count of, respectively, Caroline and Bob.

Puzzle 40 *Anne, Bob, and Caroline execute Protocol 5. What is the probability that Bob or Caroline announces before Anne that everybody has been interrogated?*

Uniform Roles In the protocols so far, different prisoners perform different roles and that was the key to solving the puzzle. Let us assume again that the prisoners know that the light is initially off. There is a protocol wherein all prisoners play the same role, but it is probabilistic, i.e., it is not determined what a prisoner will do but he has to choose between different possible actions, where we let him determine which action to choose, for example, by the outcome of throwing a couple of fair dice. This protocol is easier to present in terms of *tokens*:

Imagine each prisoner to hold a *token* worth a variable number of *points*, initially one. Turning the light on if it is off means dropping one point. Leaving the light on if it is on means not being able to drop one point. (Before, only a non-counter could drop a point.) Turning the light off if it is on, means collecting one point. Leaving the light off if it is off, means not being able to collect one point. (Before, only the counter collects points.) Protocols for n prisoners terminate once a prisoner has n points. The expression $[0, 1]$ below stands for the interval of real numbers between (and including) 0 and 1.

Protocol 6 *Entering the interrogation room, consider the number of points you carry. If the light is on, add one. Let m be this number. Let a function Pr : $\{0, ..., n\} \rightarrow [0, 1]$ be given, with $Pr(0) = Pr(1) = 1$, $0 < Pr(x) < 1$ for $x \neq 0, 1, n$, and $Pr(n) = 0$. You drop your point with probability $Pr(m)$, otherwise you collect it. The protocol terminates once a prisoner has collected n points.*

Dropping a point if you do not carry one, means doing nothing: therefore also $Pr(0) = 1$. Under the above conditions, the protocol terminates. You can get better odds than any nonzero probability if you use a function Pr that is decreasing for a prisoner holding (roughly) at most half of the number of points and is 0 if a prisoner has more than half the number of points. In other words, the first prisoner who knows that more than half of all the prisoners have been interrogated, will only act as a counter from then on (probability 0 to drop a point), and no longer as a non-counter (collecting a point). The following puzzle is an example. As $Pr(3) = 0$ in the puzzle, this does not strictly satisfy the conditions of Protocol 6, wherein $Pr(3) > 0$. We use the version with better odds.

Puzzle 41 *Show that the following interrogation sequence involving four prisoners Anne, Bob, Caroline, and Dick, where they use Protocol 6 with $Pr(0) = Pr(1) = 1$, $Pr(2) = 0.5$, $Pr(3) = 0$, and $Pr(4) = 0$, ends by Bob announcing that all prisoners have been interrogated.*

Anne, Bob, Caroline, Dick, Bob, Caroline, Caroline, Bob, Caroline, Bob

Optimization In the solution to the riddle, Protocol 4, how long does it take on average before Anne announces that everybody has been interrogated? The question is meaningless, as no interval between interrogations is known. So, how many interrogations are needed on average before Anne announces that everybody has been interrogated? That question makes sense. Of course, this depends on the scheduling of interrogations by the prison guards. Let us assume that the scheduling is random. Then, we can determine how many interrogations this takes. The answer looks nice if we assume that there is a single interrogation everyday. We do not give it here, to not to spoil the fun of trying.

Puzzle 42 *Suppose there is a single interrogation per day. How long does it take on average before the 100 prisoners can go free?*

Synchronization If there is a single interrogation everyday, it may take quite a while before the prisoners can go free. The question then comes up if they can be smarter than that and find faster protocols. If nothing is known about the interval between interrogations, we do not know of a faster protocol. But if prisoners know that there is a single interrogation per day, what can be called the case of *synchronization*, they can find faster protocols. For example, suppose that there are three prisoners, and that counter Anne is not interrogated on the first day. Then she knows immediately, even without having been interrogated herself that either Bob or Caroline must have been interrogated and she also knows that the light is now on. This sort of information can be used in protocols. We consider again the situation of 100 prisoners.

Protocol 7 *The protocol consists of two phases. The first phase takes* 100 *days and is as follows. The first prisoner who is interrogated twice turns on the light. Let us suppose this is on day m. On day* 100 *of the first phase of the protocol: if the light is off, the prisoner then interrogated announces that everybody has been interrogated; otherwise, if the light is on, the prisoner then interrogated turns off the light. The second phase is as follows; there are now three different roles. (i) The prisoner who in the first round was the first to be interrogated twice will take on the role of the counter. The counter behaves as follows: If you are being interrogated and the light is off, do nothing; if you are being interrogated and the light is on, turn it off—and keep track of how often you do this; if you have turned it off* $100 - (m - 1)$ *times, then announce that all* 100 *prisoners have been interrogated. (ii) The prisoners who have seen the light off in the first phase, and who are not the counter, do nothing in the second phase. (iii) The other prisoners take on this role: If you are being interrogated and the light is off and you have not turned it on before, then turn it on; if you are being interrogated and the light is off and you have turned it on before, then do nothing; if you are being interrogated and the light is on, then do nothing.*

If $m = 2$, then the counter has to count to 99 again, as before. Because then, the designated counter knows that nobody but himself has been interrogated when he was interrogated the second time. And in that case we do not gain time, but lose time, namely, we lose the 100 days of the first phase of execution. For m larger than 2, we can expect to gain time compared to Protocol 4. For example, if $m = 3$, then the designated counter knows that one other prisoner already has been interrogated. The (expected) loss of 100 days in phase one is now counteracted by the (expected) gain of 100 days in phase two for not having to count that prisoner plus $\frac{100}{99}$ days (i.e., 1 day) because that prisoner now does not have to turn on the light. The average number of days for someone to be interrogated twice is 13. The person then interrogated learns that 11 other prisoners must already have been interrogated. Those 11 will not do anything in the second phase, we can take them for granted. But then the counter only has to count until 88 in the second phase, instead of to 99. This shaves off about 4 years of the expected runtime before termination: after a prisoner has turned on the light, the counter has on average to wait 100 days before he is being interrogated again, and 11 times 100, plus a little bit for the noninteracting prisoners (role (ii)), is about 4 years.

9.8   History

An IBM Research webpage http://domino.watson.ibm.com/Comm/wwwr_ponder.nsf/challenges/July2002.html (from 2002) says that "this puzzle has been making the rounds of Hungarian mathematicians' parties" in a version for 23 prisoners. Apparently the riddle circulated in the USA from 2001 onward. William Wu, at the time a PhD student at Stanford University, was involved in this circle from then onward, see http://wuriddles.com. The riddle is treated in depth in the journal Mathematical Intelligencer, by Dehaye, Ford, and Segerman (2003), and also in a book entitled *Mathematical Puzzles: A Connoisseur's Collection* by Winkler (2004), in the (Dutch language) journal *Nieuwe Wiskrant* by van Ditmarsch (2007), and as *100 prisoners and a light bulb—logic and computation* by van Ditmarsch, van Eijck, and Wu (2010a) (with a protocol verification exercise in van Ditmarsch, van Eijck, and Wu (2010b)). Protocol 6, found in van Ditmarsch et al. (2010a), is by Paul-Olivier Dehaye.

Under the assumption of a single interrogation per day, better optimizations are possible than the one reported in Protocol 7. It is unknown what the smallest expected runtime is for termination. The record is about 9 years (see Dehaye et al. (2003) and http://wuriddles.com). That is already an enormous improvement over the expected duration of Protocol 4! The faster protocols distinguish even more than two roles for the prisoners, and more stages in the execution of the protocol, wherein prisoners can change role again depending on the outcome of their performance in the previous stage of execution.

10
Gossip

Six friends each know a secret. They can call each other. In each call they exchange all the secrets they know. How many calls are needed for everyone to know all secrets?

10.1 Gossip Protocols

We solve this for n friends, and let us work our way upwards bottom-up. For one friend, no calls need to be made, and for two friends, a single call between them is sufficient. For three friends, any two need to call each other first; then, either of those should call the friend who did not make a call yet; finally, the friend not involved in the second call, calls either of those involved in that call. That makes three calls.

Let there now be four friends Amal, Bharat, Chandra, and Devi (a, b, c, d) who hold, respectively, secrets A, B, C, and D. We treat a secret as a proposition with a truth-value. In that interpretation, if you know a secret, this means that you know the value of the proposition: you know whether it is true, i.e., you know that it is true or you know that it is false. So, "Amal knows secret A" means that Amal knows whether A, i.e., Amal knows that A is false or Amal knows that A is true. We represent by ab a call from a to b. The informative consequences of a call (i.e., which secrets are exchanged) are independent of who initiates a call, so in that sense a call ab is the same as a call ba. But for the generating protocols the order makes a difference. In our examples we tend to respect lexicographic order for convenience of presentation.

Back now to Amal, Bharat, Chandra, and Devi. The single call ab is sufficient to Amal and Bharat to learn their secrets, and the three calls ab; bc; ac are sufficient for Amal, Bharat, and Chandra to learn their secrets. The four calls ab; cd; ac; bd distribute all secrets over all four friends. The underlying protocol, of which this call sequence is an execution, is as follows:

Protocol 8 (Four Friends) *Any two friends make the first call; the second call is between the remaining two friends; the third call is between a friend who made the first call and a friend who made the second call; and the fourth call is between the two who were not chosen in the third call.*

The distribution of secrets given the four calls is as follows. The rows list the distribution of secrets after a particular call took place.

	a	b	c	d
	A	B	C	D
ab	AB	AB	C	D
cd	AB	AB	CD	CD
ac	ABCD	AB	ABCD	CD
bd	ABCD	ABCD	ABCD	ABCD

No other protocol solves this in four calls, and less than four calls is insufficient to distribute all secrets. We can easily show this. In an execution of any other protocol, after the first call, one of the first callers will also make the second call. So, it has to start like this.

	a	b	c	d
	A	B	C	D
ab	AB	AB	C	D
ac	ABC	AB	ABC	D
	· · ·			

How will this continue? For the third call, let us distinguish between the case that Devi (*d*) is not involved and the case that she is involved. If Devi is not involved, then another call *ac* does not result in more information, and the cases *ab* and *bc* are symmetric. Take the first, then we get

	a	b	c	d
	A	B	C	D
ab	AB	AB	C	D
ac	ABC	AB	ABC	D
ab	ABC	ABC	ABC	D
	· · ·			

Devi now has to make three more calls in order for all friends to know all secrets: one by one, she has to call Amal, Bharat, and Chandra. That makes *six* calls altogether.

Even if the third call involves Devi, there also will always remain two friends who do not know *D* yet. Again, two or three further calls are needed, so that we need at least *five* calls altogether.

Therefore: (i) less than four calls is not possible, (ii) there is no other protocol consisting of four calls, (iii) there are several call sequences such that more than four calls are needed until all friends know all secrets. In modulo symmetry (a role change of friends), any execution starts with either ab; ac (at least five calls to termination) or ab; cd (at least four calls to termination).

For $n = 4$, $2n - 4 = 4$ calls are sufficient to distribute all the secrets. Let there now be $n > 4$ friends. Also then, $2n - 4$ calls are sufficient. Suppose the friends are $a, b, c, d, e, f, \ldots$. The following protocol contains an execution sequence ab; cd; ac; bd of the Four Friends protocol. In fact, any execution of Protocol 8 would be appropriate.

Protocol 9 (Fixed Schedule) *Select four friends, and select one among those, suppose these are a, b, c, d, and a. First, a makes a call to all friends e, f, ... except b, c, d. Then, the calls ab; cd; ac; bd are made. Finally a makes, again, a call to all friends e, f, ... except b, c, d.*

This adds up to $(n - 4) + 4 + (n - 4) = 2n - 4$ calls. It will be clear that all secrets are then distributed over all friends.

Let us do this for $n = 6$, such that we get $2n - 4 = 8$ calls. Given are six friends Amal, Bharat, Chandra, Devi, Ekram, and Falguni (a, b, c, d, e, f) who hold secrets A, B, C, D, E, F. Amal starts by calling Ekram and then Falguni, etc.

	a	b	c	d	e	f
	A	B	C	D	E	F
ae	AE	B	C	D	AE	F
af	AEF	B	C	D	AE	AEF
ab	$ABEF$	$ABEF$	C	D	AE	AEF
cd	$ABEF$	$ABEF$	CD	CD	AE	AEF
ac	$ABCDEF$	$ABEF$	$ABCDEF$	CD	AE	AEF
bd	$ABCDEF$	$ABCDEF$	$ABCDEF$	$ABCDEF$	AE	AEF
ae	$ABCDEF$	$ABCDEF$	$ABCDEF$	$ABCDEF$	$ABCDEF$	AEF
af	$ABCDEF$	$ABCDEF$	$ABCDEF$	$ABCDEF$	$ABCDEF$	$ABCDEF$

This is not the only protocol to distribute the secrets in $2n - 4$ calls. For example, in Protocol 9 some calls are made more than once. For the depicted $n = 6$ execution, these calls are ae and af. The following also achieves distribution of all secrets over all friends but in *all different calls*.

	a	*b*	*c*	*d*	*e*	*f*
	A	B	C	D	E	F
ab	AB	AB	C	D	E	F
cd	AB	AB	CD	CD	E	F
ef	AB	AB	CD	CD	EF	EF
ac	ABCD	AB	ABCD	CD	EF	EF
de	ABCD	AB	ABCD	CDEF	CDEF	EF
af	ABCDEF	AB	ABCD	CDEF	CDEF	ABCDEF
bd	ABCDEF	ABCDEF	ABCD	ABCDEF	CDEF	ABCDEF
ce	ABCDEF	ABCDEF	ABCDEF	ABCDEF	ABCDEF	ABCDEF

Of course, not all sequences of eight different calls distribute the secrets over all friends. For example, when we change the sixth call above from *af* to *bf*, Amal will only know the secrets *A, B, C, D* after those eight calls.

Less than $2n - 4$ calls are insufficient to distribute all secrets for $n \geq 4$. It is not easy to prove this.

Puzzle 43 *If gossip is the goal, prolonging gossip is better! The friends do not really want to exchange their secrets as* **fast** *as possible, but as* **slow** *as possible. Well, by repeating calls in which the callers do not learn anything new, you can surely delay the moment that all secrets have been distributed. But it is boring to call the same friend without at least learning a new secret or having to tell a new secret.*

As long as two friends who call each other still exchange all the secrets that they know and at least one of them learns something new from the call, what is the maximum *number of calls to distribute all secrets?*

10.2 How to Know Whom to Call

So far, we assumed that the friends can coordinate their actions before making any calls. In the Fixed Schedule protocol for four friends, first Amal calls Bharat, and then Chandra calls Devi, and so on. The way we defined the protocol was that any two friends can make the first call, and then any two other friends can make the second call. So it does not have to be between these exact four individuals. Can the protocol be rephrased such that these scheduling decisions can be made by the friends themselves, when they are about to call someone, and based on what they know? It turns out that this is *not* possible.

One can still imagine the first two callers be determined randomly. Everybody is dying to make a call, one of the friends is simply getting through before the others in making a call, the recipient of that call can be any other friend. We then assume that the information exchange takes place instantly without consuming time and we can schedule the next call. All friends only know their own secret initially.

But for the second call we now have a problem. After the first call there are two friends who know two secrets, and the remaining friends only know one secret. In other words, they have different knowledge. In order to attempt to copy the Fixed Schedule protocol, we may pick any friend who only knows one secret to initiate the second call; this choice is knowledge-based (and anyone fulfilling the condition can be chosen), and this rules out those who made the first call. This friend now has to call another friend who only knows one secret. But the friend initiating that second call cannot choose such a one-secret-only friend *based on his knowledge*. If Chandra initiates the second call, and if she is ignorant about who made the first call, then she has no reason to prefer Devi over any other friend. It seems not unreasonable to assume that she only knows that she was not involved herself in that first call. This means that, from Chandra's point of view, the first call could have been between Amal and Bharat, or between Amal and Devi, or between Bharat and Devi. She does not know which call really happened! A second call between Chandra and Devi has to be "fixed" by an external scheduler (or by the friends themselves, prior to executing the protocol); it cannot be chosen by the friends themselves based on what they learn from receiving calls.

We conclude that it is not possible for the friends themselves to enforce the execution of the Fixed Schedule protocol.

Let us now consider an epistemic protocol wherein a friend calls another friend depending on its knowledge only, and such that any friend fulfilling the knowledge condition is chosen at random.

Protocol 10 (Learn New Secrets) *Until all friends know all secrets: choose a friend who does not know all secrets, let this friend choose a friend whose secret he does not know, and make that call.*

It is easy to see that this protocol will achieve the epistemic goal that everybody knows every secret. No call sequence obtained from the Fixed Schedule protocol can be obtained by the Learn New Secrets protocol, because in the final two calls *(ae; af)* from the Fixed Schedule protocol, Amal contacts friends of which he already knows the secret; and for the same reason Ekram and Falguni cannot call Amal again. But the same information transitions can be achieved by an execution sequence of Learn New Secrets: instead of final calls *ae; ef*, make final calls from Ekram and Falguni to Bharat (or to Chandra, or Devi): *eb; fb*. This is legal, as Ekram and Falguni do not know Bharat's secret at the time of that call. The Learn New Secrets protocol also allows for longer execution sequences than the Fixed Schedule protocol. The longest possible execution of length $n \cdot (n - 1)/2$ mentioned in Puzzle 43 is also a possible execution sequence of Learn New Secrets. For example, for $n = 4$, we get ab; ac; ad; bc; bd; cd. One can easily show that any length of call sequence between the minimum of $2n - 4$ and the maximum of $n \cdot (n - 1)/2$ can be realized by the Learn New Secrets protocol.

Puzzle 44 *What is the expected number of calls in the Learn New Secrets protocol if there are three friends? (This is easy.) If there are four friends, show that the expected number of calls is more than five. (This is not hard, but it is not as easy as for three friends.)*

10.3 Knowledge and Gossip

What sort of knowledge do the friends obtain in these protocols? This becomes interesting if we do not only consider what friends know about the secrets but also consider what they know about each other. We review what friends know in the initial state of information (wherein every friend only knows its own secret), the change of knowledge due to a call between two friends, and the knowledge they obtain after termination of a protocol consisting of such calls.

We can represent the uncertainty of the friends about their secrets in a structure. We consider Amal's secret *(A)* as a proposition of which the value is initially only known to Amal *(a)*. For four friends, a nice depiction would already be a four-dimensional structure, so let us depict the one for three friends. Below, a node like 011 stands for "*A* is false and *B* is true and *C* is true." The digits 0 and 1 stand for the truth-value of the propositions *A*, *B*, *C*, in that order.

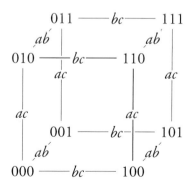

States connected with an *a*-label, or via a path consisting of *a*-labeled connections, are indistinguishable for Amal, and similarly for Bharat and Chandra.

A friend knows a proposition if and only if the proposition is true in all states indistinguishable for that friend. For example, in state 011 we have that Amal knows that A is false, because A is false in 000, 001, 010, and 011, the four states considered possible by a, and we further have that in 011, Bharat knows that B is true and Chandra knows that C is true.

The distributions of secrets over friends that we already considered in the previous section correspond in a precise way to such a structure. We represent the distribution of secrets over friends as a list (or, if you want, as a function from friends to subsets of the set of all secrets). The one above is succinctly represented by $A.B.C$, wherein Amal only knows the secret A, Bharat only knows the secret B, and Chandra only knows the secret C. The situation $AB.AB.C$, wherein Amal and Bharat both know the secrets A and B, is represented as

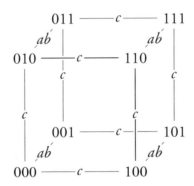

As we are only interested in whether friends know secrets and not in the truth-value of these secrets, and as any friend knows the same number of secrets in any state of such a structure, we never need to reason from the perspective of a given state, but only from the global perspective of the entire structure. For example "Amal knows whether A and whether B" is true in all states of the above structure. Therefore, we can use the convenient shorthand representation for them. We call a list like $A.B.C$ and $AB.AB.C$ a gossip state. In a gossip state, the friends have common knowledge of the distribution of secrets, i.e., each friend knows for all friends how many secrets those friends know, and it knows its own secrets. (And they all know that they all know that, etc.)

Let us now execute a telephone call in this setting. We get from $A.B.C$ to $AB.AB.C$ by executing the call ab. What sort of dynamics is a telephone call? A telephone call is a very different form of communication than an announcement in the presence of other friends. An announcement is *public*. This means that, after Amal *says* "The old name of Chennai is Madras" in the presence of Bharat and Chandra, then Bharat knows that the old name of Chennai is Madras, but Chandra also knows that Bharat knows that, and Amal knows that Chandra knows that Bharat knows that, and so on. The information that the old name of Chennai is Madras is common knowledge between the three friends. But if Amal *calls* Bharat to tell him that, and then Amal calls Chandra, and then Bharat calls Chandra, then all three know that the old name of Chennai is Madras. But still, this is not common knowledge. For example, at this stage, Amal does not know that Bharat knows it. It is even impossible that this becomes common knowledge if nothing is known about the timing of the phone calls.

We now depict the informative consequence of the sequence of calls *ab*; *ac*; *bc*, where we assume that all friends know which two friends are making a call, and when.

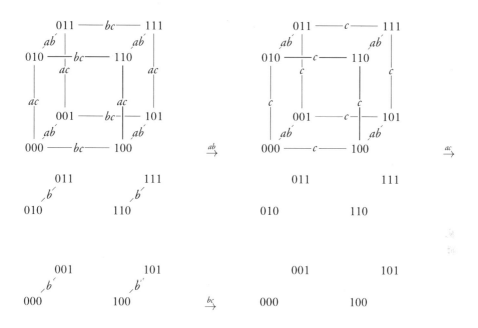

The corresponding transitions between the gossip states (the list of who knows what secrets) are as follows:

$$A.B.C \overset{ab}{\to} AB.AB.C \overset{ac}{\to} ABC.AB.ABC \overset{bc}{\to} ABC.ABC.ABC$$

Now here is an obvious but surprising observation. Having first explained that calls do not create common knowledge of the secrets, after all, at the end, there is common knowledge that all three friends know all the secrets, because in every single of the eight states, no friends consider any other state possible (and they all know this, and know that they know this, and . . .). This is because in this modeling we have assumed that the friends have common knowledge of what protocol is being carried out: they know who makes what call, even if the friend not involved cannot hear what information is being exchanged, and they know when the calls take place: the so-called assumption of synchronization.

The example above can be seen as an execution of Fixed Schedule but also as an execution of Learn New Secrets. Fixed Schedule is obvious, but Learn New Secrets is also possible on assumption of synchronization. If there are three friends and Amal and Bharat make a call, then Chandra knows that this call is taking place because she is not involved in the call. The friends (commonly) know that after three steps all friends know all secrets.

For any Fixed Schedule protocol, under conditions of synchronicity, it remains the case for more than three friends that when all friends know all secrets, they also have common knowledge of this. We could imagine the friends sitting around a table and making the "calls" from there, in view of each other, in the form of whispering, so that any other person only notices that a communication is made, but not what is being communicated. For a truly knowledge-based protocol it is no longer the case that everybody knows what call is made. We now get to that.

Consider the Learn New Secrets protocol for four friends. Suppose that Amal calls Bharat. The setting is now that each of the four friends is in their own home, out of view, sitting in front of their telephone, and in view of a telephone switchboard revealing if a call is taking place (but not revealing who the callers are). Chandra sees the switchboard light up but her phone remains dead. Similarly, it happens for Devi. Chandra and Devi now consider any call possible that does not involve them. Chandra considers it possible that the call taking place is between Amal and Bharat, or between Amal and Devi, or between Bharat and Devi. Devi considers it possible that the call taking place is between Amal and Bharat, or between Amal and Chandra, or between Bharat and Chandra. Although the real transition is $A.B.C.D \overset{ab}{\to} AB.AB.C.D$, Chandra considers it possible that the transition was $A.B.C.D \overset{ad}{\to} AD.B.C.AD$, or that it was $A.B.C.D \overset{bd}{\to} A.BD.C.BD$; Devi considers it possible that the call was one of $A.B.C.D \overset{ab}{\to} AB.AB.C.D$, $A.B.C.D \overset{ac}{\to} AC.B.AC.D$, and $A.B.C.D \overset{bc}{\to} A.BC.BC.D$. From Chandra's point of view, instead of *ab*, *ad*, or *bd*, the calls could as well have been *ba*, *da*, or *db*. If we are only interested in what secrets are learnt from the call, we abstract from who initiates the call and who receives it, so *ab* and *ba* are treated on a par, and *da* and *ad*, etc. This is also reflected by English usage: we use the word "between" when writing about two friends making a call. We now get the transition resulting in the following structure.

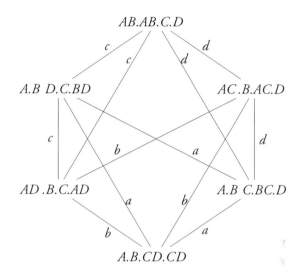

We call this structure a gossip model, with a designated gossip state for what really happened, namely *AB.AB.C.D*. As each gossip state corresponds to a 16-state structure, wherein all the values for all four secrets are combined (just like the 8-state structure for three friends and three secrets), we could even view this structure as consisting of not merely 6 but $6 \cdot 16$ states. But that representation would not have much added value.

Let us review again what the difference is between shared knowledge and common knowledge, and how this occurs in protocols and affects their execution length. We have *shared knowledge* if everybody knows something, whereas we have *common knowledge* if everybody knows something, and everybody knows that, and everybody knows that everybody knows that, and so on. Clearly the friends have *shared knowledge of all secrets* after the termination of an execution sequence of a Fixed Schedule protocol or Learn New Secrets protocol: every friend knows the value of all secrets. If they only know what calls they have received, not much more can be said. They know that the friend involved in the last call they made also knows all secrets. If we suppose that a Fixed Schedule protocol is known to the friends, then a little bit more is known: the last friends to call now know that everybody knows all secrets (and not just them). We can also achieve that everybody knows that everybody knows all secrets: let subsequently to the last call, a friend involved in that call, call all other friends. For example, following *ae; af; ab; cd; ac; bd; ae; af,* let *a* now call everybody else once more, in five calls *ab, ac, ad, ae, af*. On the assumption that this expanded protocol is known to be executed by all agents, we have now obtained shared knowledge of shared knowledge of all secrets. So then, instead of $2n - 4$ calls to achieve shared knowledge of the secrets, we need $2n - 4 + (n - 1) = 3n - 5$ calls to achieve shared knowledge of shared knowledge of the secrets, and so on, by Amal making yet another $n - 1$ calls .... . Still, it remains out of reach to make this common knowledge.

More knowledge can be achieved if we additionally assume that calls take place at regular intervals: the aforementioned condition of synchronicity, like one call every 10 minutes, and that this is also known to the friends. The secrets are then *common knowledge* after termination of a Fixed Schedule protocol! We already saw this for three friends, where it took half an hour (three calls). For six friends, this will take 1 hour and 20 minutes.

Again, we reach another threshold. Take the Learn New Secrets protocol, and four friends. We have seen that the executions consist of between four and six calls. What if general knowledge is already obtained after four calls, i.e., 40 minutes? The two friends not involved in the fourth call do not necessarily know that. But if they wait 20 more agonizing minutes, then after all, after 1 hour, it is common knowledge that everybody knows all the secrets, as no execution of the protocol takes more than six calls!

Puzzle 45 *Take an execution of the Learn New Secrets protocol of length four (everybody knows all secrets after four calls): ab; cd; ac; bd. Show that after ten more minutes everybody knows that everybody knows all secrets.*

10.4 Versions

There are various other knowledge-based gossip protocols apart from the Learn New Secrets protocol. We recall that in the Fixed Schedule protocol execution *ae; af; ab; cd; ac; bd; ae; af,* in the last two calls Amal already knows the secrets of Ekram and Falguni. But he still calls them: there is meaningful exchange of information, as in the mean time Amal has learnt secrets that Ekram and Falguni do not know yet. Consider this variation on the Learn New Secrets protocol: you call someone because you *know* that either you or the friend you call will learn a new secret. Another variation is when you call someone because you *consider it possible* that either you or the friend you call will learn a new secret. (You execute this protocol when calling your teenage daughter every hour when she is on vacation. Something might have happened over the past hour!) For that, consider the initial call sequence *ab; cd.* At this stage, Amal considers it possible that Bharat was involved in the second call and, therefore, Amal considers it possible to learn something new by calling Bharat again in the third call. Unfortunately, in the sequence ab; cd; ab, Amal and Bharat learned nothing new in the third call. But if the sequence had been ab; bc; ab, Amal would have learnt something new: C.

Puzzle 46 *The protocol where you may call another friend in case you consider it possible that he or she learnt something new, has executions that do not terminate. Give an example of a nonterminating execution sequence.*

Instead of consecutive telephone calls wherein all secrets are exchanged between both parties, several calls between two friends might as well take place at the same time to speed up the exchange of information. A number of simultaneous telephone calls is called a *round*. The question then comes up is: what the minimum number of rounds is to communicate all secrets between n friends?

Puzzle 47 *Let there be $n = 2^m$ friends. Show that m rounds suffice for all friends to get to know all secrets.*

If the number of friends is not a power of 2, we take the smallest power of 2 larger than or equal to the number of friends. Let this be m. Then, if the number of friends is even we still can do the job in m rounds, but if the number of friends is uneven we need $m + 1$ calls, as shown in the next puzzle, by example.

Puzzle 48 *Let there be five friends. Give a four-round call sequence (with parallel calls) after which all friends know all secrets.*

10.5 History

Gossip protocols have a long history in computer science. That $2n - 4$ calls is the minimum to distribute all secrets is shown by Tijdeman (1971) and in various contemporary publications; see the survey by Hedetniemi et al. (1988). The minimum number of rounds (of parallel calls) is proved by Knödel (1975) in a one-page journal article of supreme elegance. An accessible survey is by Hurkens (2000), which emerged after the riddle appeared in the 1999 NWO (Netherlands Organization for Scientific Research) Annual Science Quiz. This encouraged Hans van Ditmarsch to put a logical analysis of gossip (van Ditmarsch 2000, Section 6.6). A subsequent logical analysis is by Sietsma (2012). Knowledge-based versions of gossip protocols, such as the Learn New Secrets protocol, have been proposed by Attamah et al. (2014).

11

Cluedo

Six players are playing Cluedo. On the game board, Alice just landed in the kitchen. She says: "I suspect that Miss Scarlett did it, with a knife, in the kitchen." Nobody shows her a card. Who committed the murder?

11.1 Introduction

Cluedo (for Americans, Clue) is a murder-mystery board game wherein six partying guests are confronted with a dead body and they are all suspected of the murder. The game board depicts the different rooms of the house wherein the murder is committed, and there are also a number of possible murder weapons. There are six suspects (such as Miss Scarlett and Professor Plum), nine rooms, and six possible murder weapons. These options constitute a deck of 21 cards, one of each kind is blindly drawn and these three cards are considered the real murderer, murder weapon, and murder room. The other cards are shuffled and distributed to the players. There are usually six players. The game consists of moves that allow for the elimination of facts about card ownership, until the first player to guess the murder cards correctly has won.

The part of the game that interests us is when on the game board a room is reached by a player whose turn it is, who may then voice a *suspicion,* such as the above "I suspect that Miss Scarlett did it, with a knife, in the kitchen." A player may ask for any three cards (of the correct kinds suspect/weapon/room), also if she holds some of them herself. This question is addressed to another player and interpreted as a request to that player to admit or deny ownership of these cards. If the addressed player does not have any of the requested cards, she says so, but if a player holds at least one of the requested cards, she is obliged to show exactly one of those to the requesting player, and to that player only. The four other players cannot see which card has been shown, but of course know that it must have been one of the three.

When it is your turn, apart from voicing a suspicion you may also make an *accusation.* You may do so only once during the game. For example, your accusation may be that Miss Scarlett did it, with a knife, in the kitchen. You

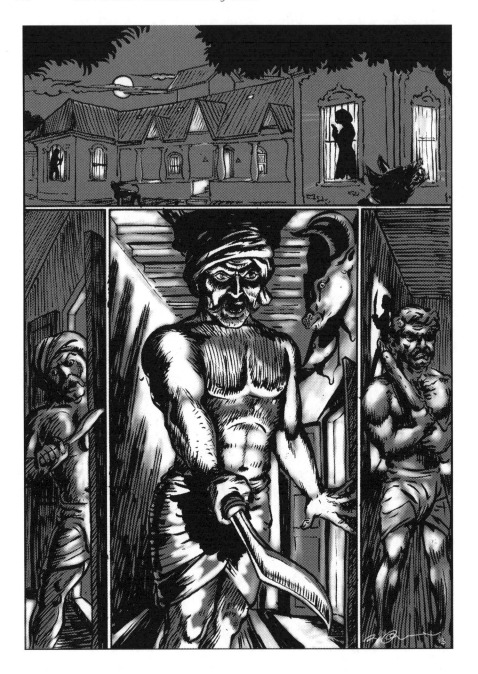

check the accusation by secretly looking at the three murder cards. If the accusation is correct, you win. If the accusation is incorrect, you only say to the other players that the accusation was incorrect and you do not tell them what the real murder cards are. The game then moves on, but you are no longer allowed to make suspicions (or an accusation). An accusation is stronger than a suspicion and fullfils a clearly different role in the game.

There are three kinds of epistemic action in Cluedo:

* A player saying that she does not have any of the three cards requested by another player.
* A player showing one of the three requested cards to another player.
* A player ending her turn without making an accusation.

We illustrate the informative consequences of these Cluedo actions by a much simpler setting. Forget about the game board and how to land in a room by throwing a pair of dice. Let there be three cards 0, 1, 2 and three players Alice, Bob, Cath (a, b, c) only. Players only know their own card. A distribution of cards is represented by a triple ijk, with $i, j, k \in \{0, 1, 2\}$ and all different. There are six different card deals. We further assume that the actual card deal is 012: Alice holds 0, Bob holds 1, and Cath holds 2. The knowledge these players initially have about their cards is represented in this model:

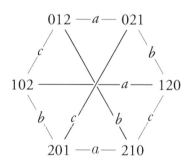

Here, for example, 012—a—021 means that Alice (a) cannot distinguish the card deal named 012, wherein she holds 0, Bob 1, and Cath 2, from the card deal 021, wherein she also holds 0 but Bob 2 and Cath 1, etc. Now, copying Cluedo, we consider three types of action: saying that you do not have a card, showing your card to another player, and saying that you cannot win, i.e., that you do not know the card deal.

11.2 I Do not Have These Cards

Suppose Alice says "I do not have card 1." As in previous chapters, we process this new information by restricting the model for the information to all states satisfying the announcement. Clearly, the only card deals not satisfying it are 120 and 102. The information transition is depicted below. From the six possible card deals, four remain. In the resulting structure, we can see that if the card deal is 012, Bob still does not know the card deal: he still cannot distinguish it from 210. But Cath, who holds 2, now knows the card deal. Her uncertainty between 012 and 102 has disappeared. More interestingly, Alice is now uncertain if Bob or Cath knows the card deal.

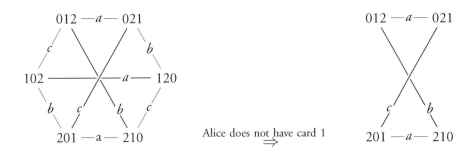

The announcement "Alice does not have card 1" (i) has the same information content as Alice *saying* "I do not have card 1," which has the same information content as the announcement "Alice *knows* that she does not have card 1" (ii). In previous chapters, it was a complication that (i) and (ii) do not always have the same information content. For example, the announcement that Bob does not have 0 is much less informative than Alice saying that Bob does not have 0: something she could only know if she were to have 0 herself!

In Cluedo, you do not merely deny ownership of a single card but of three cards. This is not greatly different from the single card situation. For example, if Alice were to say that she does not have card 1 nor card 2, we would still get a substructure from the initial hexagonal structure, namely, the one consisting of merely the following two card deals.

$$012 - a - 021$$

11.3   Showing a Card

The action of showing a card is quite different from the action of announcing that you do not have a card. Alice can only show card 0 if she actually holds card 0. But Alice can show her card, whatever card she holds. That does not eliminate any card deal from consideration. What change does it then affect? If Alice shows Bob her card, Bob will learn what Alice's card is and therefore what the card deal is. The uncertainty that Bob has between card deals should have disappeared after the action of showing a card. Alice knows that Bob will know the card deal after she has shown her card to him. But even Cath knows this without seeing the card. The effect of the action of showing a card is not a restriction of the model but what is called a *refinement* of the model: one or more players (in this case Bob) have a more refined view of the state of affairs. The informative transition is depicted below.

Alice shows her card to Bob $\Rightarrow$

Alice can only show a single card in this simplified setting. In the Cluedo action, one of the three requested cards is shown. This makes a difference, because the player showing a card may then be able to choose which card to show. We can also imagine an action with choice in the three cards example. For such an action, imagine Bob asking Alice "Please tell me a card you do not have" so that Cath hears the question, and Alice whispering the answer in Bob's ear, so that Cath cannot hear the answer, but knows that the answer has been given. In that case, whatever card Alice holds, she will always be able to choose between two answers. For example, given that she holds 0, she can whisper "I do not have 1" or "I do not have 2." As she has that option no matter what the actual card deal is, the resulting model will be the one that reflects that choice and will consist of 12 instead of 6 states. For example, we now have to distinguish the information state where the card deal is 012 and where Alice has whispered "I do not have 1" from the information state where the card deal is 012 and where Alice has whispered "I do not have 2." In the former, Bob now does not know the card deal, but in the latter, he now knows the card deal. And Cath (as she could not hear the answer) cannot distinguish between these two information states.

Similarly, in Cluedo, if Alice, Bob, and Cath are players, and if Bob utters the suspicion that Miss Scarlett did it, with a knife, in the kitchen, and Alice then shows one of her three cards to Bob, then Cath (and any other player watching the interaction) considers anything possible: Alice just showed Scarlett to Bob, or Knife, or Kitchen. Also, Alice may have more than one of these cards, in which case she can choose between different cards to show to Bob. Of course, a player watching the show action but who holds Scarlett and Knife will know by elimination that Alice must have shown the Kitchen card.

11.4   I Cannot Win

Following Alice's announcement that she does not have card 1, Bob now says "I cannot win" where this means "I do not know the card deal." In Cluedo, being able to win means that you know the murder cards, the cards on the table. You get to know these cards by finding out what cards other players hold. If you know the cards of all players, then the remaining cards must be the murder cards. (You do not always need to know the cards of all players, but merely a sufficient number in order to deduce the murder cards.) So, knowing the card deal implies knowing the murder cards, and knowing the murder cards means that you can win. By analogy, for three players each holding one card, winning means knowing what the card deal is.

From the four remaining card deals after Alice's announcement that she does not have card 1 (see below on the left), Bob knows the card deal if it is 201 and 021 and does not know the card deal if it is 012 and 210: in that case he is uncertain between 012 and 210. Therefore, Bob saying "I cannot win" rules out 012 and 210 from consideration. We get the following transition.

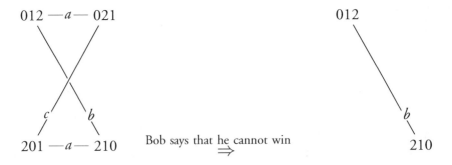

Now, surprisingly, or maybe not so surprising anymore, Alice, whose turn it now is, says "I can win!" Bob saying that he cannot win removes Alice's uncertainty between the card deals 012 and 021: in the former Bob could not win and in the latter he could; as he said he could not win, 012 must be the actual card deal. So Alice can win because Bob could not win. Cath is gnashing her teeth. Like in Cluedo, she cannot speak before her turn, but following Alice's original announcement that she does not have 1, Cath could already have won straightaway.

We finally arrive at the solution of this chapter's riddle. The riddle was:

Six players are playing Cluedo. On the game board, Alice just landed in the kitchen. She says: "I suspect that Miss Scarlett did it, with a knife, in the kitchen." Nobody shows her a card. Who did the murder?

In order to answer this question, some more precision is needed about the prior moves in the game, and we may also need to know more about the subsequent interaction between players. Let us assume that Alice landing in the kitchen was the first move in the entire game. That makes the lack of response from other players all the more surprising. Waow, a hit straightaway! This still depends. There are two continuations of this scenario.

In the first continuation, Alice now writes down her accusation "Scarlett, Knife, Kitchen," checks this with the cards on the table and triumphantly announces "I have won!" A version of this first continuation would be that she says, instead: "I lost, my accusation was incorrect," but we now rule this out, as we assume that players are perfectly rational and only make an accusation if they know it to be true. If Alice wins, it is clear why: nobody responded, so therefore no other player holds any of the three requested cards. If Alice also does not hold any of these cards, she can then deduce that the cards on the table must be Scarlett, Knife, and Kitchen, and indeed she wins.

But there is also another continuation:

Following nobody showing her a card, Alice says "OK, it's Bob's turn now, I'm done."

This does not appear to be informative at all. But with the analysis for the three card example in mind we now can conclude that Alice must hold at least one of the three requested cards. *Ending your turn in Cluedo is an announcement that you do not know the three murder cards yet!* This is always the case when you end your turn in Cluedo. If you carefully process this extra information, you may use this to your advantage and win.

For example, let us now assume Alice uttered the suspicion that Scarlett did it with a knife in the kitchen sometime later in the game, at a stage where Bob was uncertain between Alice holding the Scarlett card or Alice holding the Professor Plum card, and given that Bob already knows that two of the murder cards are Knife and Kitchen. Nobody shows a card and Alice ends her turn. (She cannot win.) Bob now concludes that Alice holds Scarlett, and that Plum is on the table. If it is now his turn, Bob will therefore immediately write down the accusation "Plum, Knife, Kitchen" and win. There is still a risk for Bob. Another explanation for her behavior is that Alice simply did not draw the correct conclusion from her suspicion and the no-show response from other players. This would have been the case if, in fact, Scarlett is on the table instead of Plum, and Alice could have concluded that Scarlett is on the table. She forgot to win. Bob will now lose by accusing Plum. It is forbidden to lie in Cluedo: you are not allowed to "forget" to show the Ballroom card if that was one of the requested cards. But it is perfectly legal to "forget" to win: not to know everything you could have known. The morale of this is that you should not count on other players being perfectly rational, even if they give fair play.

Puzzle 49 *Given card deal 012, Alice says "I do not have card 2," after which Bob says that he has won. What is the information transition resulting from Alice's announcement, and what is the information transition from Bob's subsequent announcement (i.e., saying that he knows the card deal)? What do Alice and Cath learn from Bob saying that he has won?*

11.5 How to Win Cluedo—Once

The logical analysis of the game moves in Cluedo makes it possible to compute their precise informative effects when played in a given state of the game. Given an initial deal of cards over players, this then makes it possible to build what is known as a game tree for a Cluedo game. And in principle it would then be possible to say whether "I suspect that Miss Scarlett did it with a knife in the kitchen" is a better question to ask (suspicion to utter) then "I suspect that Professor Plum did it with a rope in the kitchen." If you can reach different rooms of the house in the same move, you may even want to choose between accusations about those different rooms. But frequently you can only reach a single room. We do not know of any analysis that concludes that, in a given state of the game, from the two above suspicions the former is really better than the latter. It is also not clear in Cluedo if asking for three cards that you do not have is to be preferred over asking for three cards of which you hold or more one yourself, as in the running example of this chapter, where nobody shows a card. (Well, one thing is clear: it is unwise to ask for the three cards that you hold yourself.)

It seems complex to come up with good reasons to prefer any given suspicion over any other. Surely, for ordinary human beings playing Cluedo, this must therefore also be very complex. Is there therefore no way to win Cluedo, or at least increase your chances to win? There is: good bookkeeping. The player who remembers more than other players from past moves and their informative consequences increases his or her likelihood to win.

If you buy the board game, part of the set is a pad for detective's notes. This consists of a table where all 21 cards are listed. You can fill entries in that table if you know that a particular card is definitely the murder card or definitely not the murder card. If someone shows you the Scarlett card, you cross off Scarlett. Let us say that you then put a 0 in that entry. If in a subsequent move your suspicion is "I suspect that Professor Plum did it with a rope in the kitchen," you hold Plum and Rope yourself, and nobody is able to show you a card, then you can conclude that the murder room is Kitchen and you put a 1 in that entry. If you do not yet know whether Knife is the murder weapon, you keep that entry empty. We get a table like this. First come all the six suspect cards, then the six weapon cards, and then the nine room cards. It is exactly like the detective's notes pad, but we have only schematically represented the remaining 18 cards.

Scarlett	0
. . .	
Knife	
. . .	
Kitchen	1
. . .	

Once you have three 1 entries in the column, you are done, and you make your accusation. This can involve some logical operations (Boolean computation), for example, if you have zeros in all entries except three, then all of those must be ones.

You can do better than this by more detailed bookkeeping and more logical computations as above. You will improve your chances to win, if you do not merely keep track of what you know about the cards on the table, but also of what you know about all other cards. For that, you need a matrix with

21 rows, for the 21 cards, and 7 columns, for the 6 players $A, \dots, F$ (Alice, Bob, Chandra, Devi, Eve, Frank) and the table. You can now also record what you learn from a player showing a card to another player. Let us give a few examples. The table we had so far now looks like this. Suppose you are player A, Alice. Obviously, in the first column you only have zeros and ones. There are no empty entries in this column because you know which cards you hold yourself. We assume that apart from Plum and Rope you also hold White (i.e., Alice holds White), and that Eve holds Scarlett. If an entry is empty, you are uncertain if the player (or table) in that column holds the card in that row.

cards\owners	A	B	C	D	E	F	t
Scarlett	0	0	0	0	1	0	0
Plum	1	0	0	0	0	0	0
White	1	0	0	0	0	0	0
...	0						
Knife	0						
Rope	1	0	0	0	0	0	0
...	0						
Kitchen	0	0	0	0	0	0	1
...	0						

Suppose it is your (Alice's) move again. Alice says "I suspect that White did it with a Knife in the Ballroom." Bob says he does not have those, and Chandra shows Knife to Alice. We now get a revised table as follows.

cards\owners	A	B	C	D	E	F	t
Scarlett	0	0	0	0	1	0	0
Plum	1	0	0	0	0	0	0
White	1	0	0	0	0	0	0
...	0						
Knife	0	0	1	0	0	0	0
Rope	1	0	0	0	0	0	0
...	0						
Kitchen	0	0	0	0	0	0	1
Ballroom	0	0					
...	0						

The uncertainty about Chandra showing Knife, is to other players, not to you. But you can now also record the effect of cards shown to other players than yourself. And this exceeds the capacity of the single-column detective's notes pad. For example, it is now Bob's turn and he says "I suspect that White did it with a Candestick in the Ballroom," as a result of which Eve shows a card to Bob. And in response to a subsequent question "I suspect that Green did it with a Rope in the Ballroom" by another player, Eve shows a card to that player. We can now update as follows. We use 2s in a column for a player to indicate that the player must hold one of the cards of the rows where those 2s are. We use 3s for the same reason, and also to distinguish those entries from the entries containing 2s. The symbols need not be 2s and 3s but they could be any symbol different from the 0 and 1 that are already in use. There are two 2s and not three, because Alice knows that Eve cannot have White: she has it herself. For the same reason there are only two 3s (Alice has Rope). Once in subsequent moves, a single 2 remains in Eve's column, you (Alice) make(s) it a 1 (Eve's ownership of that card is then confirmed). Also, you then put zeros in all other entries in that entire row, as nobody else including the table now can have that card.

cards\owners	A	B	C	D	E	F	t
Scarlett	0	0	0	0	1	0	0
Plum	1	0	0	0	0	0	0
White	1	0	0	0	0	0	0
Green	0				3		
. . .	0						
Knife	0	0	1	0	0	0	0
Rope	1	0	0	0	0	0	0
Candlestick	0				2		
. . .	0						
Kitchen	0	0	0	0	0	0	1
Ballroom	0	0			23		
. . .	0						

At this stage of the game it seems pretty likely that Eve holds Ballroom. But we cannot be sure. Eve holds three cards. One of those is Scarlett. One of her two other cards is Candlestick or Ballroom, and one of her two other cards is Green or Ballroom. That is perfectly consistent with Eve having the two remaining cards Candlestick and Green. The continuation of the game will show us, by patiently plodding on, and recording the effects of moves in our expanded $21 \cdot 7$ notes pad.

We played Cluedo using such expanded notes pad a couple of times. Something funny happened and may as well happen to you. If you make notes in a table like this, you are very likely to win the game, because mere bookkeeping allows you to stay better informed than the other players. And if you are in doubt between making different suspicions, just make any suspicion. It does not matter. The bookkeeping seems to play a far bigger role than any strategic considerations. Then, the next time your friends play Cluedo with you, they all have such a notes pad and you have lost your advantage. Something else also happens: if everybody uses an expanded notes pad, all players seems to be roughly equally well informed during the game (and this is unexpected), and in the last round of the game, everybody is (or feels) very close to win the game. So the last round becomes an exciting round. (Warning: if you play this more than five times, they are no longer your friends.) If you feel that the game is almost over, it might be wise to make an accusation before you know the murder cards, for example, if you already know two but not three of the murder cards. Because if another player will make a successful accusation before it is your turn again five moves later, you will lose anyway. But that is another game, called game theory.

11.6   Versions

In the Pit game (for *trading* pit—it is a market simulation card game), the players try to corner the market in coffee, wheat, oranges, or a number of other commodities. It is like the family game in that each of these commodities are distributed over the players in the form of cards, and that players attempt to gather a full suit of cards of any commodity. A game move consists of two players trading cards. This goes about as follows. All cards you trade should be of the same suit. Players shout the number of cards they wish to exchange, simultaneously, and two players shouting the same number may then swap (trade) that number of cards (of the same suit). For example, John has two apples, three oranges, and some other cards, Mary has two oranges and yet other cards. They both shout: "Two!," "Two!," . . . , and they trade apples for oranges. John now has five oranges and Mary none, but she has two more apples. As there are nine cards of each suit, John is now closer to winning the game than before.

The trade action is somewhat similar to the action of showing a card in Cluedo: two players trading, know what cards they trade, but all other players only know that these players exchange two cards of the same suit. But Pit is also different from Cluedo, as cards actually change hands.

Puzzle 50 *Let there be three players Alice, Bob, Cath (a, b, c) and three suits Wheat, Flax, and Rye (w, x, y), of two cards each. (So, six cards altogether.) A player having two of the same suit wins. Suppose that the deal of cards is that Alice holds Wheat and Flax, Bob holds Wheat and Rye, and Cath holds Flax and Rye. This actual card deal we can represent, as in other chapters, by wx.wy.xy. What do the players know about each other's cards? That is, how many different card deals are considered possible, and by whom? You may assume that anybody holding two of a kind will already have declared a win.*

11.7 History

Cluedo was invented in 1943 by Anthony E. Pratt, a solicitor's clerk, and his wife Elva Pratt. Anthony Pratt is said to have invented the game when he was temporarily laid off because of World War II and was instead doing a, mostly boring, fire brigade duty. He had time on his hands. Elva Pratt devised the board. The Pratts' original version was called "Murder," it had ten weapons instead of six, and some suspects had other names. In the USA the game is called Clue. The logical dynamics of Cluedo are treated by van Ditmarsch (2000, 2002a). The first publication is a PhD thesis. Part of a Dutch PhD defence at the time was a "lekenpraatje" (in English: a presentation for laymen) of the research results to the general public. In that part, Hans van Ditmarsch solved the murder of Jan van Maanen by playing Cluedo with three giant-sized suspect cards (so the audience could see the cards) for the murder suspects Johan van Benthem, Gerard Renardel, and Wiebe van der Hoek. This resulted in a lot of media attention. Another Cluedo analysis is by Dixon (2006).

The Pit game was developed by Edgar Cayce in 1904. Formalizations of Pit are by Purvis et al. (2004) and by van Ditmarsch (2006).

12
Overview of Dynamic Epistemic Logic

12.1   Introduction

This is a gentle introduction to so-called dynamic epistemic logics that can describe how agents change their knowledge and beliefs. We start with a concise introduction to epistemic logic, through the example of agents holding playing cards; and, mainly for the purpose of motivating the dynamics, we also briefly introduce the concepts of shared and common knowledge. We then pay ample attention to the logic of public announcements, wherein agents change their knowledge as the result of publicly perceived events. In that setting, we also present the unsuccessful updates: formulas that become false when announced. We then present more complex epistemic updates. Finally, we briefly present a framework for jointly modeling (defeasible) belief and knowledge, and belief revision.

12.2   Epistemic Logic

We introduce epistemic logic by a simple example. Suppose there is only one agent Anne, and a stack of three playing cards.

> Anne draws one card from a stack of three different cards: clubs, hearts, and spades. Suppose that she draws the clubs card—but she does not look at her card yet; that one of the remaining cards is put back into the stack holder, suppose that is the hearts card; and that the remaining card is put (face down) on the table. That must therefore be the spades card! Anne now looks at her card.

What does Anne know? We would like to be able to evaluate system descriptions such as:

- Anne holds the clubs card.
- Anne knows that she holds the clubs card.

* Anne does not know that the hearts card is on the table.
* Anne knows her own card.

Propositions about the state of the world are in this case about card ownership. We describe such atomic propositions by, e.g., $Clubs_a$ standing for "the clubs card is held by Anne," and similarly $Clubs_h$ for "the clubs card is in the stack holder," and $Clubs_t$ for "the clubs card is on the table," etc. The standard propositional connectives are $\land$ for "and," $\lor$ for "or," $\neg$ for "not," $\rightarrow$ for "implies," and $\leftrightarrow$ for "if and only if." A formula of the form $K\varphi$ expresses that "Anne knows that φ," and a formula of the form $\hat{K}\varphi$ ($\hat{K}$ is the dual of K) expresses that "Anne can imagine that φ." The informal descriptions above become

* Anne holds the clubs card: $Clubs_a$
* Anne knows that she holds the clubs card: $KClubs_a$
* Anne does not know that the hearts card is on the table: $\neg KHearts_t$
* Anne knows her own card: $(Clubs_a \rightarrow KClubs_a) \land (Hearts_a \rightarrow KHearts_a) \land (Spades_a \rightarrow KSpades_a)$

The operator K is called a *modal* operator or a *modality*. The structures on which we will interpret descriptions using that operator are called pointed Kripke models or epistemic states (or information states). An *epistemic model* is a relational structure consisting of a *domain*, a set of "states of the world," a binary *accessibility relation* between states, and a factual description of the states—i.e., a *valuation* of atomic propositions on all states. An epistemic state is an epistemic model with a designated state. In our example, the states are card deals. The deal where Anne holds the clubs card, the hearts card is in the stack holder and the spades card is on the table, we give the name ♣♡♠, etc. By identifying states with card deals, we have implicitly specified the evaluation of atomic propositions in the state, namely, with the name ♣♡♠. The binary relation of accessibility between states expresses what the player knows about the atomic propositions. For example, if deal ♣♡♠ is actually the case, Anne holds the clubs card, and in that case she can imagine that not ♣♡♠ but ♣♠♡ is the case, wherein she also holds the clubs card. We say that state ♣♠♡ is accessible from state ♣♡♠ for Anne, or that (♣♡♠, ♣♠♡) is in the accessibility relation. Also, she can imagine the actual deal ♣♡♠ to be the case, so ♣♡♠ is "accessible from itself": the pair (♣♡♠, ♣♡♠) must also be in the accessibility relation.

Continuing in this way, we get the accessibility relation in Figure 12.1. This structure is the epistemic state ($Hexa_a$, ♣♡♠), where the epistemic model $Hexa_a = \langle S, \sim, V \rangle$ consists of a domain S, accessibility relation $\sim$, and

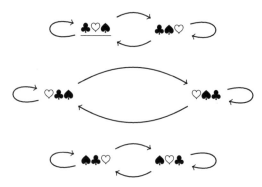

Fig. 12.1 An epistemic state that represents Anne's knowledge of the card deal where Anne holds clubs, hearts is in the stack holder, and spades is on the table. The actual state is underlined

valuation V such that

$$
\begin{aligned}
S \quad &= \quad \{\clubsuit\heartsuit\spadesuit, \clubsuit\spadesuit\heartsuit, \heartsuit\clubsuit\spadesuit, \heartsuit\spadesuit\clubsuit, \spadesuit\clubsuit\heartsuit, \spadesuit\heartsuit\clubsuit\} \\
\sim \quad &= \quad \{(\clubsuit\heartsuit\spadesuit, \clubsuit\heartsuit\spadesuit), (\clubsuit\heartsuit\spadesuit, \clubsuit\spadesuit\heartsuit), (\clubsuit\spadesuit\heartsuit, \clubsuit\spadesuit\heartsuit), \dots\} \\
V(Clubs_a) \quad &= \quad \{\clubsuit\heartsuit\spadesuit, \clubsuit\spadesuit\heartsuit\} \\
V(Hearts_a) \quad &= \quad \{\heartsuit\clubsuit\spadesuit, \heartsuit\spadesuit\clubsuit\} \\
\dots
\end{aligned}
$$

The states where a given atom is true are identified with a subset of the domain: $Clubs_a$—for "Anne holds the clubs card"—is only true in states $\{\clubsuit\heartsuit\spadesuit, \clubsuit\spadesuit\heartsuit\}$, etc. A standard modal language inductively defined by $\varphi ::= p \mid \neg\varphi \mid (\varphi \wedge \varphi) \mid K\varphi$ can now be interpreted on this structure; here p is a(n) (atomic) propositional variable and φ is a formula variable. The crucial clause in the interpretation of formulas is the one for the modal operator: $M, s \models K\varphi$ if and only if for all t, if $s \sim t$, then $M, t \models \varphi$. For $M, s \models \varphi$, read "state s of model M satisfies formula φ," or "φ is true in state s of model M." For example, we can now compute that in the epistemic state $(Hexa_a, \clubsuit\heartsuit\spadesuit)$ it is indeed true that Anne knows that she holds the clubs card:

> We have that $Hexa_a, \clubsuit\heartsuit\spadesuit \models KClubs_a$ if and only if (for all states s, if $(\clubsuit\heartsuit\spadesuit, s) \in \sim$ then $Hexa_a, s \models Clubs_a$). The last is implied by $Hexa_a, \clubsuit\heartsuit\spadesuit \models Clubs_a$ and $Hexa_a, \clubsuit\spadesuit\heartsuit \models Clubs_a$, as the only states that are accessible from $\clubsuit\heartsuit\spadesuit$ are $\clubsuit\heartsuit\spadesuit$ itself and $\clubsuit\spadesuit\heartsuit$: we have $(\clubsuit\heartsuit\spadesuit, \clubsuit\heartsuit\spadesuit) \in \sim$ and $(\clubsuit\heartsuit\spadesuit, \clubsuit\spadesuit\heartsuit) \in \sim$. Finally, $Hexa_a, \clubsuit\heartsuit\spadesuit \models Clubs_a$ because $\clubsuit\heartsuit\spadesuit \in V(Clubs_a) = \{\clubsuit\heartsuit\spadesuit, \clubsuit\spadesuit\heartsuit\}$, and, similarly, $Hexa_a, \clubsuit\spadesuit\heartsuit \models Clubs_a$ because $\clubsuit\spadesuit\heartsuit \in V(Clubs_a) = \{\clubsuit\heartsuit\spadesuit, \clubsuit\spadesuit\heartsuit\}$. Done!

Anne's accessibility relation is an equivalence relation. If one assumes what are known as the properties of knowledge, this is always the case. The properties

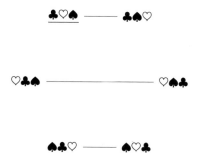

Fig. 12.2 A simpler visualization of the epistemic state where Anne holds clubs, hearts is in the stack holder, and spades is on the table. The actual state is underlined

are that "what you know is true," which is formalized by the schema $K\varphi \rightarrow \varphi$; that "you are aware of your knowledge," which is formalized by the schema $K\varphi \rightarrow KK\varphi$, and that "you are aware of your ignorance," which is formalized by the schema $\neg K\varphi \rightarrow K\neg K\varphi$. These properties may be disputed for various reasons, for example, without the requirement that what you know is true, we get a notion of belief instead of knowledge. And there are also many things you do not know without being aware of it! For the sake of a simple exposition, we will stick to the properties of knowledge and see where they get us. As said, together they enforce that in epistemic logic the accessibility relation is always an equivalence relation. This is somewhat differently expressed, by saying that what an agent cannot distinguish induces a *partition* on the set of states, i.e., a set of equivalence classes that cover the entire domain. For equivalence relations, as they are symmetric, it is customary to write them "infix," i.e., ♣♡♠ ∼ ♣♠♡ instead of (♣♡♠, ♣♠♡) ∈ ∼. In the case of equivalence relations, a simpler visualization than with arrows is sufficient: we only need to link visually the states that are in the same class. If a state is not linked to others, it must be a singleton equivalence class (reflexivity always holds). For $(Hexa_a, ♣♡♠)$ we get the visualization in Figure 12.2.

One might ask: why not restrict ourselves in the model to the two deals ♣♡♠ and ♣♠♡ only? The remaining deals are inaccessible anyway from the actual deal! From an agent's point of view this is arguably right, but from a modeler's point of view the six-point model is preferable: this model works regardless of the actual deal.

The dual of "know" is "can imagine that" (or "consider it possible that"): this modality is often defined by abbreviation as $\hat{K}\varphi := \neg K\neg\varphi$. If you can imagine that a proposition is true, you do not know that the proposition is not true. For example, "Anne can imagine that the hearts card is not on the table" is described by $\hat{K}\neg Hearts_t$ which is true in epistemic state $(Hexa_a, ♣♡♠)$, because from deal ♣♡♠ Anne can access deal ♣♠♡ for which ¬*Hearts_t* is true, as the spades card is on the table in that deal. There is no generally accepted notation for "can imagine that." Other notations for $\hat{K}\varphi$ are $M\varphi$, $L\varphi$, and $k\varphi$.

> The modality for knowledge is K, but the more general notation for modalities is $\Box$, also called the "necessity" operator. The "possibility" operator $\Diamond$ is the dual of $\Box$, i.e., $\Diamond\varphi$ is equivalent to $\neg\Box\neg\varphi$. Because the knowledge modality is interpreted on structures with an equivalence relation, we have used $\sim$ for that relation, in general one will often see R instead (for "relation," no doubt).

12.3   Multiagent Epistemic Logic

Many features of formal dynamics can be presented based on the single-agent situation. For example, the action of Anne picking up the card that has been dealt to her from the table is a significantly complex epistemic action. But a proper and more interesting perspective is the multiagent situation. This is because players may now have knowledge about each others' knowledge, so that for even a single atomic proposition the epistemic models representing that knowledge can become arbitrarily complex. Let us imagine three players Anne, Bill, and Cath, each holding one card from a stack of three (known) cards clubs, hearts, and spades, such that they know their own card but do not know which other card is held by which other player. Assume that the actual deal is that Anne holds clubs, Bill holds hearts and Cath holds spades. The epistemic operator K with corresponding access $\sim$, to describe Anne's knowledge, now has to be different from an epistemic operator and corresponding access for Bill, and yet another one for Cath. The distinction can easily be made by labeling the knowledge operator and the accessibility relation with the agent. If we take a for Anne, b for Bill, and c for Cath, this results in equivalence relations $\sim_a$, $\sim_b$, and $\sim_c$ and corresponding knowledge operators K_a, K_b, and K_c. Bill's access on the domain is different from Anne's, whereas Anne cannot tell deals ♣♡♠ and ♣♠♡ apart, Bill instead cannot tell deals ♣♡♠ and ♠♡♣ apart, etc. The epistemic state (*Hexa*, ♣♡♠) is pictured in Figure 12.3, and we can now describe in the epistemic language that:

* Anne knows that Bill knows that Cath knows her own card: $K_a K_b((Clubs_c \rightarrow K_c Clubs_c) \wedge (Hearts_c \rightarrow K_c Hearts_c) \wedge (Spades_c \rightarrow K_c Spades_c))$.
* Anne has the clubs card, and she knows that, but Anne knows that Bill can imagine that Cath knows that Anne does not have the clubs card: $K_a Clubs_a \wedge K_a \hat{K}_b K_c \neg Clubs_a$.

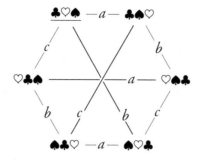

Fig. 12.3 The epistemic state (*Hexa*, ♣♡♠) for the card deal where Anne holds clubs, Bill holds hearts, and Cath holds spades

The structures we will use throughout this presentation can now be introduced formally as follows:

Definition 1 (Epistemic structures) An *epistemic model* $M = \langle S, \sim, V \rangle$ consists of a *domain S* of (factual) *states* (or "worlds"), an *accessibility* function $\sim : A \to \mathcal{P}(S \times S)$, and a *valuation* $V : P \to \mathcal{P}(S)$. For $s \in S$, (M, s) is an *epistemic state*.

For $\sim (a)$ we write $\sim_a$ and for $V(p)$ we write V_p. So, the accessibility function $\sim$ can be seen as a set of equivalence relations $\sim_a$, and V as a set of valuations V_p. We often omit the parentheses in (M, s). Relative to a set of agents A and a set of atoms P, the language of multiagent epistemic logic is inductively defined by $\varphi ::= p \mid \neg\varphi \mid (\varphi \wedge \varphi) \mid K_a\varphi$. (Also here, we often omit parentheses.) We need some further extensions of the language before we put it into a formal definition, but all these extensions will be interpreted on the structures presented in Definition 1.

The epistemic models now defined are of course the structures we have already seen in almost all previous chapters. The main difference is that we used informal English descriptions of knowledge and ignorance there, to be interpreted on these structures, instead of the formal counterparts, formulas in the logical language. Propositions like "Anne holds clubs" are now represented by propositional variables $Clubs_a$. In complex propositions, we use $\wedge$ instead of "and," and so on for other propositional connectivity. And instead of "knows" we write K. The proposition "Alice knows that Cath does not know any of the other players' cards" that featured in the Russian Cards problem, can now be written as $K_a((0_a \to \neg K_c 0_a) \wedge (1_a \to \neg K_c 1_a) \wedge \ldots \wedge (0_b \to \neg K_c 0_b) \wedge \ldots)$. From a certain perspective that is all the difference there is. The English descriptions are informal, but they are precise.

12.4   Common Knowledge

We can extend the logical language with epistemic operators for *groups* of agents. We will add common knowledge operators. (There are also other extensions.) As we aim to focus on *dynamic* epistemics in this contribution, and not on dynamic *epistemics*, this will be a lightningly quick introduction to "common knowledge."

In the epistemic state (*Hexa*, ♣♡♠) of Figure 12.3, both Anne and Bill know that the deal of cards is *not* ♠♣♡: both $K_a \neg (Spades_a \wedge Clubs_b \wedge Hearts_c)$ and $K_b \neg (Spades_a \wedge Clubs_b \wedge Hearts_c)$ are true. If a group of agents all individually know that φ, we say that φ is *shared knowledge*. The modal operator for shared knowledge of a group B is E_B. For an arbitrary subset $B \subseteq A$ of the set of agents A, we define $E_B \varphi := \bigwedge_{a \in B} K_a \varphi$. So in this case we have that $E_{ab} \neg (Spades_a \wedge Clubs_b \wedge Hearts_c)$—we abuse the language and write E_{ab} instead of $E_{\{a,b\}}$. Now φ may be generally known, but that does not imply that agents know about each other that they know φ. For example, $K_b K_a \neg (Spades_a \wedge Clubs_b \wedge Hearts_c)$ is *false* in (*Hexa*, ♣♡♠): Bill can imagine Anne to have spades instead of clubs. In that case, Anne can imagine that the card deal is ♠♣♡. So $\hat{K}_a \hat{K}_b (Spades_a \wedge Clubs_b \wedge Hearts_c)$ is true, and therefore $K_b K_a \neg (Spades_a \wedge Clubs_b \wedge Hearts_c)$ is false. For other examples, one can construct formulas that are true to some extent $K_a K_b K_c K_a K_a K_b \varphi$ but no longer if one adds one more operator at the start, e.g., $K_b K_a K_b K_c K_a K_a K_b \varphi$ could then be false. A formula φ is *common knowledge* for a group B, notation $C_B \varphi$, if it holds for arbitrary long stacks of individual knowledge operators (for individuals in that group). If, for example, $B = \{a, b, c\}$, we get something (involving an enumeration of all finite stacks of knowledge operators) like $C_{abc} \varphi := \varphi \wedge K_a \varphi \wedge K_b \varphi \wedge K_c \varphi \wedge K_a K_a \varphi \wedge K_a K_b \varphi \wedge K_a K_c \varphi \wedge \ldots K_a K_a K_a \varphi \ldots$. Alternatively, we may see common knowledge as the conjunction of arbitrarily many applications of general knowledge: $C_B \varphi := \varphi \wedge E_B \varphi \wedge E_B E_B \varphi \wedge \ldots$. Such infinitary definitions are frowned upon. Therefore, common knowledge C_B is added as a primitive operator to the language, whereas shared knowledge is typically defined (for a finite set of agents) by the notational abbreviation above. This does not matter, because common knowledge is defined semantically, by an operation on the accessibility relations for the individual agents in the group, namely, transitive closure of their union. By way of validities involving common knowledge that are mentioned at the end of this section, any single arbitrarily large conjunct from the right-hand side of the infinitary definition of common knowledge is then entailed.

The semantics of common knowledge formulas is: $C_B \varphi$ is true in an epistemic state (M, s) if φ is true in any state s_m *that can be reached by a finite path of* *linked states* $s \sim_{a_1} s_1 \sim_{a_2} s_2 \sim_{a_3} \cdots \sim_{a_m} s_m$, with all of $a_1, \ldots, a_m \in B$ (and not necessarily all different). Mathematically, "reachability by a finite path" is the same as "being in the transitive reflexive closure." We define $\sim_B$ as $(\bigcup_{a \in B})^*$; this is the reflexive transitive closure of the union of all accessibility relations for agents in B. Then, we interpret common knowledge as

$$M, s \models C_B \varphi \text{ if and only if for all } t : s \sim_B t \text{ implies } M, t \models \varphi$$

If all individual accessibility relations are equivalence relations, $\sim_B$ is also an equivalence relation. Common knowledge for the entire group A of agents is called *public knowledge*.

In the model *Hexa*, access for any subgroup of two players, or for all three, is the entire model. For such groups B, $C_B \varphi$ is true in an epistemic state (*Hexa*, t) iff φ is valid on the model *Hexa*—a formula is valid on a model M, notation $M \models \varphi$, if and only if for all states s in the domain of M: $M, s \models \varphi$. For example, we have that:

- It is public knowledge that Anne knows her card:
  Hexa $\models C_{abc}(K_a Clubs_a \vee K_a Hearts_a \vee K_a Spades_a)$.
- Anne and Bill share the same knowledge as Bill and Cath:
  Hexa $\models C_{ab}\varphi \rightarrow C_{bc}\varphi$.

Valid principles for common knowledge are $C_B(\varphi \rightarrow \psi) \rightarrow (C_B \varphi \rightarrow C_B \psi)$ (distribution of C_B over $\rightarrow$), and $C_B \varphi \rightarrow (\varphi \wedge E_B C_B \varphi)$ (use of C_B), and $C_B(\varphi \rightarrow E_B \varphi) \rightarrow (\varphi \rightarrow C_B \varphi)$ (induction). Some grasp of group concepts of knowledge is important to understand the effects of public announcements.

Common knowledge often featured informally in the epistemic riddles in the previous chapters. We have also stayed away from being very precise about it, as in natural language we can only approach it with the infinite iteration of shared knowledge. The semantic definition by transitive closure is more direct. In the three cards setting, common knowledge is not so interesting, as after two iterations we can reach all states in the model: if $K_a K_b \varphi$ is true (or a stack of two knowledge operators for any other two of the three agents), then $C_{abc}\varphi$ is also true! Now take, for a more meaningful example, the consecutive numbers puzzle. In the model below for Consecutive Numbers, it is common knowledge to Anne and Bill that Anne's number is odd (and in the other infinite chain, not depicted, it is common knowledge to Anne and Bill that Anne's number is even; we assume that "odd" is a propositional variable only true in states where Anne's number is odd). No finite iteration of K_a and K_b operators is able to express this information.

$$10 - a - 12 - b - 32 - a - 34 - b - \cdots$$

12.5   Public Announcements

We now move on to the dynamics of knowledge. Suppose Anne says that she does not have the hearts card. She then makes it public to all three players that all deals where $Hearts_a$ is true can be eliminated from consideration. This results in a restriction of the model *Hexa* as depicted in Figure 12.4. Anne's announcement "I do not have hearts" is interpreted as a public announcement of "Anne knows that she does not have hearts," in this case, that has the same meaning as the public announcement of "Anne does not have hearts." This can be seen as an epistemic program with precondition $\neg Hearts_a$ that is interpreted as a transformer of the original epistemic state, exactly as a program in dynamic modal logic. Given some program π, in dynamic logic, $[\pi]\psi$ means that after every execution of π (state transformation induced by π), formula ψ holds. Given some model (relational structure), every execution of π corresponds to a pair in the accessibility relation for program π in that model: the first argument in the pair is the state before execution, and the second argument in the pair is the state after the execution. For announcements, we want something of the form $[\varphi]\psi$, meaning that after (every) announcement of φ, formula ψ holds.

We appear to be moving away from the standard paradigm of modal logic. So far, the accessibility relations were between states in an epistemic model. But all of a sudden, we are confronted with an accessibility relation between epistemic states as well: "I do not have hearts" induces an epistemic state transition such that the pair of epistemic states in Figure 12.4 is in that relation. The epistemic states take the role of the points or worlds in a seemingly underspecified domain of "all possible epistemic states." It is not as bad as it seems! We do not need to refer to a vague domain of all epistemic states, but to the concrete domain of, given an epistemic model, all epistemic states such that their domains are modally definable subsets of that given epistemic model (*modally definable* means that the subset is the restriction of the domain to all states where a modal formula φ is true: the announcement). By lifting the accessibility relation between points in the original epistemic state to an accessibility relation between epistemic states, we can get the dynamic and epistemic accessibility relations on the same level again, and see this as a relational structure on which to interpret a perfectly ordinary multimodal logic. A crucial point is that this structure is induced by the initial epistemic state and the actions that can be executed there, and not the other way round. So announcement modalities are standard modalities, after all.

The announcement "Anne does not have hearts" is a simple epistemic action in various respects. It is public, and therefore not private (Anne telling Cath her card without Bill noticing) or another form of nonpublic. It is truthful.

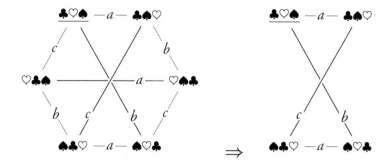

Fig. 12.4 On the left, the epistemic state (*Hexa*, ♣♡♠) for the card deal where Anne hold clubs, Bill holds hearts, and Cath holds spades. The actual deal is underlined. On the right, the effect of Anne saying that she does not have hearts

This means two things in our setting: Anne announces "I do not have hearts" because she knows that she does not have hearts or she can see the card in her hands. She is not allowed to lie. But this also means that the announcement "Anne does not have hearts" is (taken to be) true: therefore we restrict the model to states where it is true. There are many variations here that we do not discuss: there are non-truthful announcements (lying agents), and there is a semantics of announcements that only assumes that agents believe what they say, but not that announcements are true. This alternative semantics is more suitable to model belief and change of belief (instead of knowledge and change of knowledge). We all bypass that in high speed. Finally, an announcement is a special program because it is deterministic, i.e., it is a state transformer; other actions (whispering a card you do not have to another player) are nondeterministic: they can have multiple outcomes in a given state.

The effect of the public announcement of φ is the restriction of the epistemic state to all states where φ holds. So, "announce φ" can indeed be seen as an epistemic state transformer, with a corresponding dynamic modal operator $[\varphi]$. We now finally introduce the logical language with all the operators we have seen so far.

Definition 2 (Logical language of public announcements) Given are a set of agents A and a set of atoms P. Let $p \in P$, $a \in A$, and $B \subseteq A$ be arbitrary. The language of public announcements is inductively defined as

$$\varphi ::= p \mid \neg\varphi \mid (\varphi \wedge \varphi) \mid K_a\varphi \mid C_B\varphi \mid [\varphi]\varphi$$

Definition 3 (Semantics) Given is an epistemic model $M = \langle S, \sim, V \rangle$, and $s \in S$. We define:

$$
\begin{array}{lll}
M, s \models p & \text{iff} & s \in V_p \\
M, s \models \neg\varphi & \text{iff} & M, s \not\models \varphi \\
M, s \models \varphi \wedge \psi & \text{iff} & M, s \models \varphi \text{ and } M, s \models \psi \\
M, s \models K_a\varphi & \text{iff} & \text{for all } t \in S, s \sim_a t \text{ implies } M, t \models \varphi \\
M, s \models C_B\varphi & \text{iff} & \text{for all } t \in S, s \sim_B t \text{ implies } M, t \models \varphi \\
M, s \models [\varphi]\psi & \text{iff} & M, s \models \varphi \text{ implies } M|\varphi, s \models \psi
\end{array}
$$

where $M|\varphi = \langle S', \sim', V \rangle$ is defined as follows:

$$
\begin{array}{lll}
S' & = & \{s' \in S \mid M, s' \models \varphi\} \\
\sim'_a & = & \sim_a \cap (S' \times S') \\
V'_p & = & V_p \cap S'.
\end{array}
$$

The model $M|\varphi$ is the model M restricted to all the states where φ holds (including a restriction of the accessibility relation between states). The interpretation of the dual $\langle\varphi\rangle$ of $[\varphi]$ is:

$$M, s \models \langle\varphi\rangle\psi \text{ if and only if } M, s \models \varphi \text{ and } M|\varphi, s \models \psi.$$

Formula φ is valid on model M, notation $M \models \varphi$, if and only if for all states s in the domain of M: $M, s \models \varphi$. Formula φ is valid, notation $\models \varphi$, if and only if for all models M (of the class of models for the given parameters of A and P): $M \models \varphi$.

For an example, we can now verify with a semantic computation, that after Anne's announcement that she does not have hearts, Cath knows that Anne has clubs (see Figure 12.4).

In order to prove that $Hexa, \clubsuit\heartsuit\spadesuit \models [\neg Hearts_a]K_cClubs_a$, we have to show that $Hexa, \clubsuit\heartsuit\spadesuit \models \neg Hearts_a$ implies $Hexa|\neg Hearts_a, \clubsuit\heartsuit\spadesuit \models K_cClubs_a$. As it is indeed the case that $Hexa, \clubsuit\heartsuit\spadesuit \models \neg Hearts_a$ (as $\clubsuit\heartsuit\spadesuit \notin V_{Hearts_a} = \{\heartsuit\clubsuit\spadesuit, \heartsuit\spadesuit\clubsuit\}$), it only remains to show that $Hexa|\neg Hearts_a, \clubsuit\heartsuit\spadesuit \models K_cClubs_a$. The set of states in the model $Hexa|\neg Hearts_a$ that is equivalent to $\clubsuit\heartsuit\spadesuit$ for Cath is the singleton set $\{\clubsuit\heartsuit\spadesuit\}$. Therefore, it is sufficient to show that $Hexa|\neg Hearts_a, \clubsuit\heartsuit\spadesuit \models Clubs_a$, which follows trivially from $\clubsuit\heartsuit\spadesuit \in V_{Clubs_a} = \{\clubsuit\heartsuit\spadesuit, \clubsuit\spadesuit\heartsuit\}$.

The semantics of public announcement (that we have given here as it is usually given) is slightly imprecise. Consider what happens if in "$M, s \models [\varphi]\psi$ if and only if $M, s \models \varphi$ implies $M|\varphi, s \models \psi$" the formula φ is false in M, s. In that case, $M|\varphi, s \models \psi$ is undefined, because s is now not a part of the domain of the model $M|\varphi$. Apparently, we informally use that an implication is true not only when the antecedent is false and the consequent true or false, but also that an implication is true when the antecedent is false and the consequent undefined. A more precise definition of the semantics of public announcement that does not have that informality is: $M, s \models [\varphi]\psi$ if and only if for all (M', t) such that $(M, s)i\varphi(M', t)$: $(M', t) \models \psi$, where $i\varphi$ is a binary relation, in infix notation, between epistemic states. In this definition, $(M, s)i\varphi(M', t)$ holds if and only if $M' = M|\varphi$ and $s = t$.

To give the reader a feel for what goes in this logic we give some of its valid principles. In all cases we only give motivation and we refrain from proofs.

If an announcement can be executed, there is only one way to do it:

$$\langle\varphi\rangle\psi \rightarrow [\varphi]\psi \text{ is valid.}$$

This is a simple consequence of the functionality of the state transition semantics for the announcement. The converse $[\varphi]\psi \rightarrow \langle\varphi\rangle\psi$ does not hold. Take $\varphi = \psi = \bot$ ($\bot$ is "falsum"). We now have that $[\bot]\bot$ is valid but $\langle\bot\rangle\bot$ is, of course, always false, because no epistemic state satisfies $\bot$. All the following are equivalent:

* $\varphi \rightarrow [\varphi]\psi$
* $\varphi \rightarrow \langle\varphi\rangle\psi$
* $[\varphi]\psi$

A sequence of two announcements can always be replaced by a single, more complex announcement. Instead of first saying "φ" and then saying "ψ" you may as well have said for the first time "φ and after that ψ." It is expressed by

$$[\varphi \wedge [\varphi]\psi]\chi \text{ is equivalent to } [\varphi][\psi]\chi.$$

This validity is a useful feature for analyzing announcements that are made with intentions, or other conversational implicatures. Intentions can sometimes be modeled as postconditions ψ that should hold after the announcement. The announcement of φ with the intention of achieving ψ is in that case really an announcement of $\varphi \wedge [\varphi]\psi$.

For an example sequence, consider the following announcement made by an outsider that has full knowledge of the epistemic state (*Hexa*, ♣♡♠).

An outsider says: "The deal of cards is neither ♠♣♡ nor ♡♠♣."

This is formalized as $\neg(Spades_a \wedge Clubs_b \wedge Hearts_c) \wedge \neg(Hearts_a \wedge Spades_b \wedge Clubs_c)$. Abbreviate this announcement as one. Figure 12.5 depicts the result of the announcement of one. Observe that none of the three players Anne, Bill, and Cath know the card deal as a result of this announcement! Now imagine that the players know (have common knowledge) that the outsider made the announcement one in the happy knowledge of not revealing the deal of cards

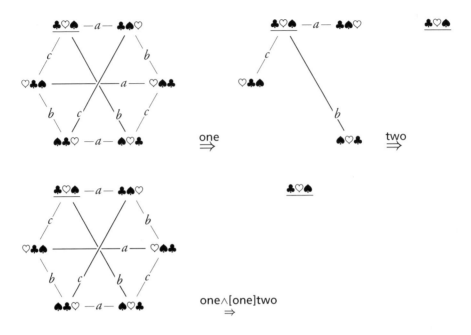

Fig. 12.5 A sequence of two announcements can be replaced by a single announcement

to anyone. For example, he might have been boasting about his logical prowess and the players might inadvertently have become aware of that. In other words, it becomes known that the announcement one was made with the intention of keeping the players ignorant of the card deal. Ignorance of the card deal (whatever the deal may have been) can be described as some long formula that is a conjunction of 18 parts and that starts as $\neg K_a(Clubs_a \wedge Hearts_b \wedge Spades_c) \wedge \neg K_b(Clubs_a \wedge Hearts_b \wedge Spades_c) \wedge \neg K_c(Clubs_a \wedge Hearts_b \wedge Spades_c) \wedge \ldots$ and that we abbreviate as two. The formula two is false in all states that are a singleton equivalence class for at least one player, and true anywhere else. So in the model *Hexa*|one resulting from the announcement of one, formula two is only true in state ♣♡♠. For the result of the announcement of two, see again Figure 12.5. Observe that in the epistemic state resulting from two, all players now know the card deal. So in that epistemic state, two is false. Now what does it mean that the players have become aware of the intention of the outsider? This means that although the outsider was actually saying one, he really meant "one, and after that two," in other words, he was saying one ∧ [one] two. Unfortunately, we have that *Hexa*, ♣♡♠ ⊨ [one ∧ [one] two]¬ two. The outsider could have kept the card deal a secret, but by intending to keep it a secret he was revealing the secret.

In $[\varphi]\psi$, the logical relation of the announced formula φ to the postcondition ψ of the announcement is not trivial. Combining two announcement by a single announcement is already an example of that, it comes with the validity $[\varphi][\psi]\chi \leftrightarrow [\varphi \wedge [\varphi]\psi]\chi$. But other cases are also interesting.

We now consider the case where the postcondition is a knowledge formula, i.e., a formula of the form $K_a\psi$. Then $[\varphi]K_a\psi$ is not equivalent to $K_a[\varphi]\psi$, because the interpretation of the modality $[\varphi]$ is a partial function between epistemic states. A simple counterexample is the following: in $(Hexa, \clubsuit\heartsuit\spadesuit)$, where Anne holds clubs, it is true that after every truthful announcement of Anne holding hearts, Cath knows that Anne holds clubs. This is because the announcement cannot be made in truth. In other words, we have

$$Hexa, \clubsuit\heartsuit\spadesuit \models [Hearts_a]K_cClubs_a.$$

On the other hand, it is false that Cath knows that after the announcement of Anne that she holds the hearts card, Anne holds the clubs card. This is because Cath considers it possible that Anne holds hearts, and if this announcement were truthfully made, then after the announcement it would still be true that Anne holds hearts, so that she does not hold clubs. So we have

$$Hexa, \clubsuit\heartsuit\spadesuit \not\models K_c[Hearts_a]Clubs_a.$$

If we make $[\varphi]K_a\psi$ conditional to the truth of the announcement, an equivalence holds:

$$[\varphi]K_a\psi \text{ is equivalent to } \varphi \to K_a[\varphi]\psi.$$

For negation, we also get an equivalence that is conditional to the executability of the announcement: another schematic validity of the logic is $[\varphi]\neg\psi \leftrightarrow (\varphi \to \neg[\varphi]\psi)$. Now if we consider two more, and list them all, except for the case "common knowledge," we get this:

$$
\begin{array}{lcl}
[\varphi]p & \leftrightarrow & (\varphi \to p) \\
[\varphi](\psi \wedge \chi) & \leftrightarrow & [\varphi]\psi \wedge [\varphi]\chi \\
[\varphi]\neg\psi & \leftrightarrow & (\varphi \to \neg[\varphi]\psi) \\
[\varphi]K_a\psi & \leftrightarrow & \varphi \to K_a[\varphi]\psi \\
[\varphi][\psi]\chi & \leftrightarrow & [\varphi \wedge [\varphi]\psi]\chi.
\end{array}
$$

This is useful. With the exception of the last one, on the left-hand side, the announcement binds another logical operator, and on the right-hand side it is pushed beyond that operator. This provides a recipe to eliminate all public announcement operators from a formula by rewriting it into an equivalent formula. For example:

$$[\neg Hearts_a]K_c\neg Hearts_a \leftrightarrow \neg Hearts_a \to K_c[\neg Hearts_a]\neg Hearts_a$$
$$\leftrightarrow \neg Hearts_a \to K_c(\neg Hearts_a \to \neg[\neg Hearts_a]Hearts_a)$$
$$\leftrightarrow \neg Hearts_a \to K_c(\neg Hearts_a \to \neg(\neg Hearts_a \to Hearts_a))$$
$$\leftrightarrow \neg Hearts_a \to K_c(\neg Hearts_a \to \neg Hearts_a)$$
$$\leftrightarrow \neg Hearts_a \to K_c\top$$
$$\leftrightarrow \top.$$

So, in this case the equivalent formula is the trivial formula $\top$ ("always true"). We can do this for every formula. The trivial formula will not always result, but indeed a formula without announcements. The case of a formula with two subsequent announcements, as in the last equivalence above, seems bothersome. But it is not really a bother: if we employ an inside-out rewriting strategy, and also use the principle that equivalent subformulas of a given formula can be substituted for one another, then we will still always succeed. An alternative way to proceed is not to use that principle (as it needs proof) but to use a complexity measure on formulas wherein form $[\varphi][\psi]\chi$ is more complex than form $[\varphi \wedge [\varphi]\psi]\chi$.

Because every formula in public announcement logic is logically equivalent to one in the logic without announcements, multiagent epistemic logic, this shows that public announcement logic without common knowledge has the same what is known as *expressive power* as multiagent epistemic logic (i.e., it can define the same properties on the set of epistemic states). And we thus also obtain the complete axiomatization of the logic, a systematic way to derive all validities of the logic. We will not delve into such matters.

Common knowledge was left out for a good reason: if we add common knowledge, it is no longer the case that every formula with public announcements is equivalent to one without announcements. The straightforward generalization of the principle $[\varphi]K_a\psi \leftrightarrow (\varphi \to K_a[\varphi]\psi)$ relating announcement and individual knowledge to $[\varphi]C_A\psi \leftrightarrow (\varphi \to C_A[\varphi]\psi)$ is invalid.

Consider a model M for two agents a and b and two atomic propositions p and q. Its domain is $\{11, 01, 10\}$, where 11 is the state where p and q are both true, 01 the state where p is false and q is true, and 10 the state where p is true and q is false. Agent a cannot tell 11 and 01 apart, whereas b cannot tell 01 and 10 apart. So, the partition for a on the domain is $\{11, 01\}, \{10\}$ and the partition for b on the domain is $\{11\}, \{01, 10\}$.

$$10 \;-\!b\!-\; 01 \;-\!a\!-\; 11 \qquad \overset{p}{\Rightarrow} \qquad 10 \qquad\qquad 11$$

Now consider the formulas $[p]C_{ab}q$ and $(p \rightarrow C_{ab}[p]q)$. We show that the first is true in state 11 of M, whereas the second is false there. First, $M, 11 \models [p]C_{ab}q$ is true in 11, because $M, 11 \models p$ and $M|p, 11 \models C_{ab}q$. For the result of the announcement of p in $(M, 11)$, see above. The model $M|p$ consists of two disconnected states; obviously, $M|p, 11 \models C_{ab}q$, because $M|p, 11 \models q$ and 11 is now the only reachable state from 11. On the other hand, we have that $M, 11 \not\models p \rightarrow C_{ab}[p]q$, because $M, 11 \models p$ but $M, 11 \not\models C_{ab}[p]q$. The last is because $11 \sim_{ab} 10$ (because $11 \sim_a 01$ and $01 \sim_b 10$), and $M, 10 \not\models [p]q$. When evaluating q in $M|p$, we are now in the *other* disconnected part of $M|p$. But there q is false: $M|q, 10 \not\models q$.

Finding the right principle for the interaction of common knowledge and announcement was part of the history of dynamic epistemic logic. There are two solutions to that quest.

In the first place, in order to get a complete axiomatization of the logic there is a rule relating validities: "If $\chi \rightarrow [\varphi]\psi$ and $\chi \wedge \varphi \rightarrow E_A\chi$ are valid, then $\chi \rightarrow [\varphi]C_A\psi$ is valid." The counterpart of that in the axiomatization is what is known as a *derivation rule*, stating that if instantiations of two premises (such as axioms) have already been derived, then the conclusion of form $\chi \rightarrow [\varphi]C_A\psi$ is also derivable.

But we can also extend the language even more, namely, with operators for *conditional common knowledge* $C_B^\psi \varphi$ (also called *relativized* common knowledge), meaning that the agents in group B have common knowledge of φ *on condition ψ*. We recall the semantics of $C_B\varphi$: this is true in a state s, if φ is true in all states t that can be reached by a finite path built from accessibility links for agents in B. Conditional common knowledge $C_B^\psi \varphi$ is true in a state s, if φ is true in all states t that can be reached by a finite path built from accessibility links for agents in B and such that condition ψ is satisfied in every state on that path. Ordinary common knowledge $C_B\varphi$ is then definable as $C_B^\top \varphi$. Conditional common knowledge $C_B^\psi \varphi$ is not the same as $C_B(\varphi \wedge \psi)$: clearly, in the state at the end of a finite ψ-path φ and ψ are both true. But there may also be paths at the end of which φ and ψ are true, but such that ψ is not true along the way all the time.

Then, after all, an axiom instead of a derivation rule for such common knowledge is possible, namely, $[\varphi]C_B^\psi \chi \leftrightarrow (\varphi \rightarrow C_B^{\varphi \wedge [\varphi]\psi}[\varphi]\chi)$, of which an instantiation for $\psi = \top$ is $[\varphi]C_B\chi \leftrightarrow (\varphi \rightarrow C_B^\varphi[\varphi]\chi)$. We will not get into details here, but explain this by the example model M above: here we have that $[p]C_{ab}q$ is equivalent to $p \rightarrow C_{ab}^p[p]q$. The latter is true in state 11, because the state 10 is now unreachable, as we cannot bridge the gap caused by state 01, where condition p is false.

12.6   Unsuccessful Updates

After announcing φ, φ may remain true but may also have become false! This is puzzling (and therefore you are holding this book with knowledge puzzles). To understand why, we go back into the history of modal logic, beyond the modern dynamic connotations. The formula $p \wedge \neg Kp$ is known as a Moore-sentence, after the British moral philosopher G. E. Moore. Moore-sentences cannot be known. In other words, $K(p \wedge \neg Kp)$ is inconsistent in epistemic logic. This can easily be seen by the following argument: from $K(p \wedge \neg Kp)$ follows $Kp \wedge K\neg Kp$, so follows Kp. But from $Kp \wedge K\neg Kp$ also follows $K\neg Kp$, and from that follows, using the properties of knowledge (and of belief), that $\neg Kp$. Together, Kp and $\neg Kp$ are inconsistent.

In dynamic epistemic logic, this gets a different, dynamic setting: I can tell you that p is true and that you do not know that. After which, you know that p, so that this announced sentence $p \wedge \neg Kp$, has become false: $\neg(p \wedge \neg Kp)$ is equivalent to $\neg p \vee Kp$ which is entailed by Kp. This is not problematic or impossible, it is merely an (admittedly crucial) observation about dynamics. In dynamic epistemic logic, this sort of announcement has been called an *unsuccessful update*: a true announcement is made such that the announcement formula is false afterwards. If the goal of the announcing person was to "spread the truth of this formula," then this attempt was clearly unsuccessful.

We appear to be deceived by some intuitive, but incorrect, communicative expectation. If a true announcement φ is made to an agent, it appears on first sight that this announcement of φ makes φ known to the agent: in other words, if φ is true, then after the announcement $K\varphi$ is true (where K describes the knowledge of the agent—to make our point, it suffices to consider a single, unlabeled, agent). In other words, $\varphi \rightarrow [\varphi]K\varphi$ appears to be valid. This expectation is unwarranted, because the truth of epistemic parts of the formula may be influenced by the announcement of the formula. But on the other hand—it is not that our intuition is *that* stupid—this expectation is sometimes warranted. Quite a few formulas always become known after being announced. These can be called successful. Let us begin with the simplest possible example.

$$ 0 \underline{\hspace{1.5cm}} \underline{1} \quad \overset{p \wedge \neg Kp}{\Longrightarrow} \quad \underline{1} $$

There is one atomic proposition p, and one (anonymous) agent, and an epistemic model M encoding that the agent is uncertain about p. It consists of states 0 where p is false and 1 where p is true, and these states are indistinguishable for the agent. The actual state is 1. The announcement of $p \wedge \neg Kp$ restricts the model to the states where it is true. Well, p is only true in state 1. But $\neg Kp$ is also true in 1, because the agent considers 0 possible wherein p is false. So,

$p \wedge \neg Kp$ is true in 1, and obviously false in 0. The model restriction therefore consists of state 1 only. In this singleton model $M|(p \wedge \neg Kp)$, the agent knows that p: Kp is true. But (as said above), if Kp is true, then also $\neg p \vee Kp$, which is equivalent to $\neg(p \wedge \neg Kp)$, the negation of the announcement. So $p \wedge \neg Kp$ is false after its announcement. If the announcement formula is false, then it is also not known (where we use the dual of the principle of knowledge that anything known is true). But in this case this is evident, as you cannot know $p \wedge \neg Kp$ anyway. Winding up the results, we now have

$$
\begin{aligned}
M, 1 &\models p \wedge \neg Kp \\
M|(p \wedge \neg Kp), 1 &\models \neg(p \wedge \neg Kp) \\
M, 1 &\not\models \langle p \wedge \neg Kp \rangle (p \wedge \neg Kp) \\
M, 1 &\not\models \langle p \wedge \neg Kp \rangle K(p \wedge \neg Kp) \\
M, 1 &\not\models (p \wedge \neg Kp) \to [p \wedge \neg Kp] K(p \wedge \neg Kp) \\
&\not\models (p \wedge \neg Kp) \to [p \wedge \neg Kp] K(p \wedge \neg Kp).
\end{aligned}
$$

So for $\varphi = p \wedge \neg Kp$, we do not have $\models \varphi \to [\varphi] K\varphi$.

Let us continue with another example, of a more involved multi-agent character. Consider Anne announcing in the epistemic state $(Hexa, \clubsuit\heartsuit\spadesuit)$: "Bill does not know that I hold clubs." By conversational implicature, this affirms the truth of that fact, so that it means "Anne holds clubs and Bill does not know that Anne holds clubs," and, as Anne is saying it and only says what she knows to be true, we get "Anne knows that: Anne holds clubs and Bill does not know that Anne holds clubs." This is the announcement of $K_a(Clubs_a \wedge \neg K_b Clubs_a)$. After this announcement, Bill now knows that Anne holds the clubs card, so $K_b Clubs_a$ has become true, and therefore $\neg(Clubs_a \wedge \neg K_b Clubs_a)$ as well, and thus also $\neg K_a(Clubs_a \wedge \neg K_b Clubs_a)$. The reader can simply check in Figure 12.6 that after its announcement, the formula of the announcement has become false. In a multiagent setting, the communicative expectation is that the announcement formula becomes *common knowledge*, and an unsuccessful formula fails to achieve that. We indeed also have the required $Hexa, \clubsuit\heartsuit\spadesuit \not\models \langle K_a(Clubs_a \wedge \neg K_b Clubs_a) \rangle C_{abc} K_a(Clubs_a \wedge \neg K_b Clubs_a)$, so that $\not\models K_a(Clubs_a \wedge \neg K_b Clubs_a) \to [K_a(Clubs_a \wedge \neg K_b Clubs_a)] C_{abc} K_a(Clubs_a \wedge \neg K_b Clubs_a)$. So we do not have $\models \varphi \to [\varphi] C_A \varphi$ either.

The two examples we have now seen are formulas that, once announced, remain false forever. They are always unsuccessful, so to say. There are also formulas that are always successful, such as the plain announcement of p in $(M, 1)$ above, or Anne saying that she holds clubs in the other example, without bothering to nag Bill about his ignorance. If a true fact is announced, it is always common knowledge afterwards: $\models p \to [p] C_A p$. In between these extremes of "always successful" and "always unsuccessful," there are also formulas that sometimes remain true, and at other times—given other epistemic states—become false after an announcement.

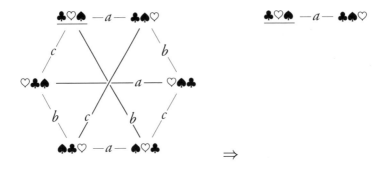

Fig. 12.6 Anne says to Bill: "You do not know that I hold clubs."

A typical example is "not stepping forward" in the Muddy Children problem of Chapter 3. Let there be three children, and let m_i, for $i = a, b, c$, stand for "child i is muddy." The first announcement in the Muddy Children problem is father saying "At least one of you is muddy." This is easy: $m_a \vee m_b \vee m_c$. We abbreviate this formula as one. Then, father says "If you know whether you are muddy, please step forward." We recall that the "announcement" (the publicly observed event) in Muddy Children is the simultaneous response of the children to father's request to step forward. If the response is that nobody steps forward, this actually means "nobody knows whether he/she is muddy." For example, "Anne knows whether she is muddy" is formalized by "$K_a m_a \vee K_a \neg m_a$," so that "nobody knows whether he/she is muddy" is formalized by

$$\neg(K_a m_a \vee K_a \neg m_a) \wedge \neg(K_b m_b \vee K_b \neg m_b) \wedge \neg(K_c m_c \vee K_c \neg m_c).$$

Call this formula nostep (for "not stepping forward"). The negation $\neg$nostep is true when at least one child knows whether he/she is muddy, so this formula is true when the muddy children step forward. Below we depict once more the information transitions for the problem.

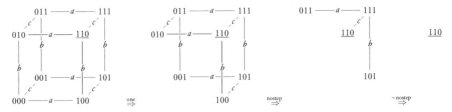

The formula expressing that no child knows whether it is muddy, is true at first, becomes false when announced (when nobody steps forward), and then remains false when its negation is announced (when Alice and Bob step forward). Now consider that all three children are muddy. Then father has to repeat his request three times. The first time nobody steps forward, it remains true that no child knows whether it is muddy. The second time, this becomes false. So we have

$$M, 111 \models \text{one}$$
$$M|\text{one}, 111 \models \text{nostep}$$
$$M|\text{one}|\text{nostep}, 111 \models \text{nostep}$$
$$M|\text{one}|\text{nostep}|\text{nostep}, 111 \models \neg\text{nostep}.$$

The formula nostep is sometimes successful, and sometimes unsuccessful. Now consider n children, of which k are muddy, and let nostep_n formalize that n children do not know whether they are muddy (so that the above nostep is nostep_3). Father then has to repeat his request k times. The first $k-2$ times the children's response (the announcement nostep_n) is successful, the $(k-1)$st time it is unsuccessful (the formula $\neg\text{nostep}_n$ is now true), and therefore the kth time they step forward.

The following terminology describes all those nuances.

Definition 4 (Successful formula/Successful update) A formula φ in the language of public announcements is *successful* if and only if $[\varphi]\varphi$ is valid. A formula is *unsuccessful* if and only if it is not successful. Given an epistemic state, (M, s), φ is a *successful update* in (M, s) if and only if $M, s \models \langle\varphi\rangle\varphi$; and φ is an *unsuccessful update* in (M, s) if and only if $M, s \models \langle\varphi\rangle\neg\varphi$.

We recall that $\langle\varphi\rangle$ is the dual of $[\varphi]$: $\langle\varphi\rangle\psi$ means by abbreviation $\neg[\varphi]\neg\psi$; alternatively, given the announcement properties, we can see $\langle\varphi\rangle\psi$ as $\varphi \land [\varphi]\psi$.

Announcements of successful formulas are always successful updates, but sometimes successful updates are on formulas that are unsuccessful. The intuitive meaning of "unsuccessful" is a relation between an epistemic state and a formula, not a property of a formula. Calling a formula unsuccessful has therefore the drawback that all inconsistent formulas are successful.

We can link our intuitions about "success" to the definition of a successful formula in an elegant way: A formula $[\varphi]\varphi$ is valid, if and only if $[\varphi]C_A\varphi$ is valid, if and only if $\varphi \to [\varphi]C_A\varphi$ is valid. So the successful formulas do what we want them to do: if true, they become common knowledge when announced. It is not known what formulas are successful! For a single agent, this question has been answered (and the answer is highly technical), but not for multiple agents. An answer to this question is not obvious, because even if φ and ψ are successful, $\neg\varphi$, $\varphi \land \psi$, or $\varphi \to \psi$ may be unsuccessful. For example, both p and $\neg Kp$ are successful formulas, but, as we have seen, $p \wedge \neg Kp$ is not.

> It is puzzling when a formula becomes false because it is announced. That is clearly the reason that epistemic puzzles are called *puzzles*. Almost all riddles treated in this book concern announcements that become false, or ignorance turning into knowledge. In the Consecutive Numbers puzzle, Anne and Bill get to know the other's number by both of them saying that they do not know it. We have to tailor the problem to our modeling needs: "Anne does not know Bill's number" is a successful update, and "Bill does not know Anne's number" is also a successful update. But "Anne does not know Bill's number, and after that Bill does not know Anne's number" is an unsuccessful update. In the Hangman, the surprise is spoilt by being announced. In Russian Cards, the intention to guard a secret makes you leak information and lose the secret. In Sum and Product, S and P initially do not know the number pair but learn it from their announcements about each other's ignorance. Careful modeling, like we did above in detail for Muddy Children, and like we sketched for Consecutive Numbers, allows to construct unsuccessful updates in these settings.

12.7   Epistemic Actions

The effect of a public announcement is a restriction of the epistemic model. Some epistemic actions are not public, and then the effect of the action is not a restriction of the epistemic model. Let us reconsider the epistemic state (*Hexa*, ♣♡♠) for three players Anne, Bill, and Cath, each holding one of clubs, hearts, and spades; and wherein Anne holds clubs, Bill holds hearts, and Cath holds spades. Consider the following action (it has been treated in detail in Chapter 11, for other cards).

> Anne shows (only) to Bill her clubs card. Cath cannot see the face of the shown card, but notices that a card is being shown.

As always in this epistemic setting, it is assumed that it is publicly known what the players can and cannot see or hear. Call this action showclubs. The epistemic state transition induced by this action is depicted in Figure 12.7. Unlike after public announcements, in the showclubs action we cannot eliminate any state. Anne can show her card whatever the card deal is. Instead, all b-links between states have now been severed. Whatever the actual deal of cards, Bill will know that card deal after Anne's card showing action. The reason that no card deals are eliminated is that no card is publicly known not to have been

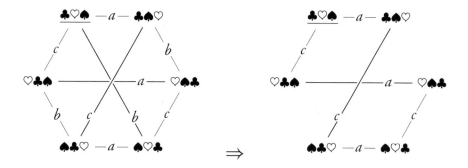

Fig. 12.7 On the *left*, the epistemic model for three players each holding one card. On the *right*, the effect of Anne showing her clubs card to Bill

shown. Let us explain. The way to model these epistemic actions is to consider all that can take place from the perspective of any agent, where one should also take into account what agents consider possible about other agents.

* Clearly, Anne showing clubs cannot be ruled out, because she actually shows clubs.
* Anne showing hearts can also not be ruled out. Anne can imagine that Cath can imagine that Anne has hearts, because Anne can imagine Cath to have spades, and so not to know whether Anne will show clubs or hearts; so it might have been hearts.
* Anne showing spades can also not be ruled out. Anne can imagine Cath not to have spades but hearts instead, in which case Cath would not have known whether Anne has shown clubs or spades; so it might have been spades.

Bill's and Cath's perspective on the epistemic action is rather different. The moment Bill sees Anne's card, he will rule out any other action he considered possible prior to the execution. Before the action, he thought he would see clubs or spades. But he did see (and now knows that Anne holds) clubs. Cath can only rule out that Anne showed spades, because Cath has spades herself. Cath considers it possible that Anne showed clubs or hearts. But Anne does not know that Cath can rule out spades, because even after the action Anne still considers it possible that Cath holds hearts. What counts in modeling is this higher-order perspective. From that perspective, Cath can in principle (i.e., if we do not know what her actual card is) not distinguish between any of the three showing actions.

We can think of the action showclubs as a structured epistemic action, relating the three possible actions wherein Anne shows clubs, hearts, and spades such that Anne and Bill are known to be able to distinguish between the three, and such that Cath cannot distinguish between any of them. The preconditions for the three show actions are that Anne actually holds the card that she is showing. What results is an epistemic action that is like an epistemic model, a relational structure consisting of a domain and an indistinguishability relation for each agent, except that instead of valuations of atomic propositions in each state we now have preconditions associated to each action. The action that really happened (showing clubs) is designated (in the figure, underlined). As the obvious names of the three actions, we have chosen ♣, ♡, and ♠, where $\text{pre}(♣) = Clubs_a$, etc.

The next question then becomes how, given the initial epistemic state (*Hexa*, ♣♡♠), one constructs the epistemic state wherein Bill always knows the card deal from executing this epistemic action showclubs. That is the final part in this story on the development of dynamic epistemic logic. We proceed with sketching what is known as *action model logic*.

An *action model* is a structure like epistemic model but with a precondition function instead of a valuation function.

Definition 5 (Action model) An *action model* $M = \langle S, \approx, \text{pre} \rangle$ consists of a *domain* S of *actions*, an *accessibility function* $\approx: A \rightarrow \mathcal{P}(S \times S)$, where each $\approx_a$ is an accessibility relation, and a *precondition function* $\text{pre} : S \rightarrow \mathcal{L}$, where $\mathcal{L}$ is a logical language. A pointed action model is an *epistemic action*.

In this overview, we only consider action models with accessibility relations that are equivalence relations, but there is no general restriction of that kind. (For example, to model change of belief instead of change of knowledge, we consider relations that are not equivalence relations.) A truthful public announcement of φ is a singleton action model with precondition φ and with the single action accessible to all agents. Action model logic is a generalization of public announcement logic.

Performing an epistemic action in an epistemic state means computing what is known as their restricted modal product. This product encodes the new state of information.

Definition 6 (Update of an epistemic state with an action model) Given an epistemic state (M, s) where $M = \langle S, \sim, V \rangle$ and an epistemic action (M, s) where $M = \langle S, \approx, \text{pre} \rangle$. Let $M, s \models \text{pre}(s)$. The update $(M \otimes M, (s, s))$ is the epistemic state where $M \otimes M =$

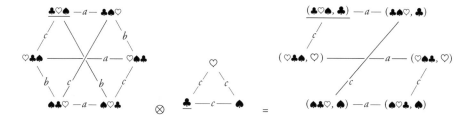

Fig. 12.8 Anne shows clubs to Bill, as the execution of an action model

$\langle S', \sim', V' \rangle$ such that

$$
\begin{array}{lll}
S' & = & \{(t, \mathsf{t}) \mid M, t \models \mathrm{pre}(\mathsf{t})\} \\
(t, \mathsf{t}) \sim'_a (t', \mathsf{t}') & \textit{iff} & t \sim_a t' \; \textit{and}\; \mathsf{t} \approx_a \mathsf{t}' \\
(t, \mathsf{t}) \in V'_p & \textit{iff} & t \in V_p.
\end{array}
$$

The domain of $M \otimes \mathsf{M}$ is the product of the domains of M and M, but restricted to state/action pairs (t, t) such that $M, t \models \mathrm{pre}(\mathsf{t})$, i.e., such that the action can be executed in that state. (A state in the resulting domain is no longer an abstract object, as before, but such a state/action-pair. We can still see it as an abstract object, modeling the operation slightly differently.) An agent cannot distinguish pair (t, t) from pair (t', t') in the next epistemic state if she cannot distinguish states t and t' in the initial epistemic state and also cannot distinguish actions t (that is executed in t) and t' (that is executed in t'). The valuations do not change after action execution. This is a logic of knowledge change, not a logic of factual change.

In the logical language, we can associate a dynamic operator to the execution of an epistemic action, very similar to the dynamic operator for the announcement: $[\mathsf{M}, \mathsf{s}]\varphi$ means that after every execution of epistemic action (M, s), φ is true. Skipping over a number of technical details, the semantics of this modality is then as follows.

$$
M, s \models [\mathsf{M}, \mathsf{s}]\varphi \quad \textit{iff} \quad M, s \models \mathrm{pre}(\mathsf{s}) \text{ implies } (M \otimes \mathsf{M}), (s, \mathsf{s}) \models \varphi.
$$

We now depict again the result of executing the epistemic action of Anne showing clubs in the epistemic state where Anne, Bill, and Cath each hold a single card, but apply the definitions above. See Figure 12.8. For example, $(\clubsuit\heartsuit\spadesuit, \clubsuit)$ is a pair in the resulting model, because $\mathrm{pre}(\clubsuit) = \textit{Clubs}_a$ and $\textit{Hexa}, \clubsuit\heartsuit\spadesuit \models \textit{Clubs}_a$. In the resulting model, we have that $(\clubsuit\heartsuit\spadesuit, \clubsuit) \sim_a (\clubsuit\spadesuit\heartsuit, \clubsuit)$, because $\clubsuit\heartsuit\spadesuit \sim_a \clubsuit\spadesuit\heartsuit$ and $\clubsuit \sim_a \clubsuit$. However, $(\clubsuit\heartsuit\spadesuit, \clubsuit) \not\sim_b (\spadesuit\heartsuit\clubsuit, \spadesuit)$, because $\clubsuit \not\sim_b \spadesuit$. Etc.

To give another example of an epistemic action, consider the following action, rather similar to the action of Anne showing her card.

Anne whispers into Bill's ear that she does not have the spades card, given a (public) request from Bill to whisper into his ear one of the cards that she does not have.

Given that Anne has clubs, she could have whispered "no hearts" or "no spades." And whatever the actual card deal was, she could always have chosen between two such options. It comes with an epistemic action (model) that is just like the one for Anne showing her card, except that the three preconditions for action execution are now $\neg Clubs_a$, $\neg Hearts_a$, and $\neg Spades_a$, respectively. We expect an epistemic state to result that reflects that choice, and that therefore consists of $6 \cdot 2 = 12$ different states. It is depicted in Figure 12.9. The reader may ascertain that the desirable postconditions of this whisper action indeed hold. For example, given that Bill holds hearts, Bill will now have learnt from Anne whispering "no spades" what Anne's card is, and thus the entire deal of cards. So there should be no alternatives for Bill in the actual state (the underlined state ♣♡♠ "at the back" of the figure—different states for the same card deal have been given the same name, as they can be distinguished by their epistemic properties, how they relate to other states). But Cath does not *know* that Bill knows the card deal, as Cath can imagine that Anne actually whispered "no hearts" instead. That would have been something that Bill already knew, as he holds hearts himself—so from that action he would not have learnt very much. Note that in Figure 12.9 there is also another state named ♣♡♠, "in the middle," so to speak, that is accessible for Cath from the state ♣♡♠ "at the back." Therefore, Cath cannot distinguish the result of Anne whispering "no spades" given that Anne has clubs, from the result of Anne whispering "no hearts" given that Anne has clubs. Cath cannot distinguish any states linked by a chain of c-links. For example, she can also not distinguish the result of Anne whispering "no hearts" given that Anne has clubs, from the result of Anne whispering "no spades" given that Anne has hearts.

There is much more to be said about this action model logic. We have not touched the issue of how action model modalities interact with other operators, in view of an axiomatization. We have not addressed common knowledge in this framework. Further generalizations are possible wherein we also allow factual change (i.e., wherein we allow to change the value of atomic propositions when performing an epistemic action). The interaction of epistemic change and factual change is interesting, and *puzzling*: we have seen this in the version of the Muddy Children problem wherein Anne gets cleaned (which results in a change of the value of the proposition "Anne is

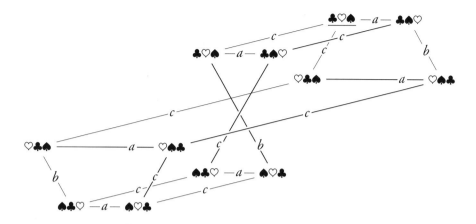

Fig. 12.9 Anne whispers to Bill that she does not have spades

muddy"), and in the "One Hundred Prisoners and a Light Bulb" riddle the light is switched on and off all the time, in combination with not publicly observable actions: only the prisoner being interrogated will see the state of the light. We will give references for further reading on action model logic and related works in the notes section at the end.

12.8 Belief Revision

In dynamic epistemic logic you cannot change your mind. Once you know a fact (an atomic proposition), you know it forever: once Kp is true, it remains true after every update. As we are modeling knowledge, this is obvious: one of the properties of knowledge is that known propositions are true ($K\varphi \rightarrow \varphi$). So, there is no need to change your mind. The theoretical setup for changing knowledge, as we presented it, is easily generalizable to other epistemic notions, such as belief. A standard difference between belief and knowledge is that beliefs may be false. Let us write $B\varphi$ for "the agent believes φ." It is possible that you believe that the card on the table is clubs, but that in fact it is spades: $BClubs \wedge \neg Clubs$ is consistent. It is not problematic to give logical principles for belief and to interpret such a modality on epistemic structures: we then have accessibility relations that are not equivalence relations. (An equivalence relation is reflexive, so that the actual state is always considered possible. But in mistaken beliefs the actual state—"what really is the case"—is not considered possible.)

What we want is a logic of change of belief wherein factual belief Bp can be changed into factual belief $B\neg p$, and that can also handle change of complex belief. In a different community, not surprisingly called that of "belief revision" and that operates in the wider area of artificial intelligence, such change of belief is the most natural operation around—the operation is then indeed called belief revision. In the development of dynamic epistemic logic, how to model belief revision came after how to model change of knowledge. In this section, we shortly survey the basic principles.

Consider one agent and an atomic proposition p that she is is uncertain about. We can explain most that is relevant for belief change by a single agent only, with an unlabeled belief operator B. Let us consider Anne, who is uncertain which of three cards is on the table: clubs, hearts, or spades. In fact, spades is on the table. The epistemic model $(T, \spadesuit)$ (T for table) coming with that is as follows, where in this case we do not assume transitivity (in view of later enrichments of this structure).

In this example, there are three atomic propositions *Clubs, Hearts, Spades*, where *Clubs* is only true in ♣, etc. So far, this is not different from what we have seen. But now it becomes different: Anne may consider some states more plausible than others. For example, she may find it more plausible that the card is clubs than that the card is hearts, and more plausible that the card is hearts than that the card is spades. She believes what she finds most plausible. Therefore, she believes that the card is clubs: *BClubs*. This is unrelated to what card is really on the table. That is spades. We then have that $T, \spadesuit \models Spades \wedge BClubs$ and also that $T, \spadesuit \models \neg Clubs \wedge BClubs$. Formally, we can model Anne's preferences among these states as a relation $<$ that consists of the pairs $(\clubsuit, \heartsuit)$, $(\clubsuit, \spadesuit)$ and $(\heartsuit, \spadesuit)$. It is common to take the reflexive closure $\leq$ of such a relation, so that $\clubsuit \leq \clubsuit$ expresses that clubs is at least as plausible as itself, and as this obviously goes in both directions, that clubs is equally plausible as itself. (More generally, we can have that different states can be equally plausible, for example, Anne may find it equally plausible that clubs and spades are on the table, and less plausible that hearts is on the table than that clubs or spades are on the table).

We can also model knowledge K in this setting. The plausibility relation $\leq$ is required to satisfy certain properties ($\leq$ should be a well-preorder: it should be reflexive and transitive, and it should have the property that every nonempty subset has an element that is at least as plausible as all the others in that set), and given that, we can define $\sim$ as the symmetric closure of $\leq$:

$s \sim t$ iff $s \leq t$ or $t \leq s$. Then, $\sim$ is an equivalence relation that can be used to interpret knowledge: $K\varphi$ is true in state s iff φ is true in all states t that are indistinguishable from s (such that $t \sim s$). For example, Anne *knows* that the card on the table is either clubs, spades, or hearts. This is the accessibility relation in the model T. If we combine $\sim$ and $>$ in one picture, we can depict this as

Now imagine that the agent wants to revise her current beliefs. Spades is on the table. Anne believes that clubs is on the table, but has been given sufficient reason to believe that this is false. She wants to incorporate the information that $\neg Clubs$. We can accomplish this by making the states where *Clubs* is false more plausible than the state where *Clubs* is true. So, change the accessibility relation $<$ by removing (♣, ♡) and (♣, ♠) and by adding (♡, ♣) and (♠, ♣). What should we do about the remaining preference for hearts over spades? A reasonable approach is to leave that unchanged. So (♡, ♠) will still be in the new plausibility relation. In our example we therefore get this revised model T':

Observe that ♡ is now the most plausible state, so that we now have that $T', ♠ \models BHearts$. The knowledge has not changed. It is still the case that $T', ♠ \models K(Clubs \lor Hearts \lor Spades)$.

This operation of belief change can be modeled in the logical language with a dynamic modality $[*\varphi]$, for "belief revision with formula φ that is interpreted with a plausibility changing operation as outlined below and as in the above example." Applied to the example, we can now say that $T, Spades \models BClubs \land [*\neg Clubs]BHearts$. In general, when revising with a formula φ, a minimal way to revise your beliefs is to make all states satisfying φ more plausible than all states not satisfying φ, but to leave all plausibilities among the φ states unchanged, and similarly to leave all plausibilities among the $\neg\varphi$ states unchanged.

Then, we can do this for multiple agents, with different belief revision policies, in the presence of common belief operators, and for nonpublic kinds of belief revision: the whole carpet of dynamic epistemic logic can be invitingly rolled out again. None of our epistemic riddles has used belief revision, they were all in terms of change of knowledge, but in the history of dynamic epistemic logic, how to model belief revision is an important topic.

12.9   Beyond Dynamic Epistemic Logic

We have seen the logic of public announcements, a logic of epistemic actions, and a bit of belief revision in dynamic epistemic logic. Much more has been done, some of what has already been mentioned. Instead of change of knowledge, we can change belief, change both at the same time (as in the previous section), or investigate the dynamics of a wealth of other epistemic notions. We can combine epistemic change (such as change of knowledge) with factual change (as in cleaning muddy children, or switching lights). Executing an action can be seen as a "tick of the clock," i.e., as a temporal step, and this lays a bridge between logics of action and knowledge and logics of time and knowledge, between dynamic epistemic logics and temporal epistemic logics. Given any of these many logics, researchers are interested in finding out if you can determine whether a formula is true or false in a given model (model checking) in finding out whether a formula has a model (satisfiability), and the computational complexity of such decision problems. Typically, axiomatizations are given: mechanical procedures to determine if formulas follow from a set of other formulas. The main problem is then to determine that such proof procedures are sound and complete: soundness means that no invalid formulas can be derived in such proofs, and completeness means that all valid formulas are found. Some references to all these topics are found in the next section with notes.

12.10   Historical Notes

Epistemic Logic and Common Knowledge The logic of knowledge as a modal logic is often attributed to Hintikka (1962), although Hintikka himself is too much of an academic and a gentleman to want to get this credit and always refers to yet older roots. His 1962 "Knowledge and Belief" makes eminent reading, also today. Common knowledge is following on the heels of knowledge and has been pioneered by Lewis (1969), Friedell (1969), Aumann (1976), and McCarthy (1990) (published notes from the late 1970s). Conditional common knowledge is a much more recent invention, by Kooi and van Benthem (2004); van Benthem et al. (2006). Excellent introductions into epistemic logic and common knowledge are by Fagin et al. (1995) and by Meyer and van der Hoek (1995), and a recent handbook is edited by van Ditmarsch et al. (2015).

Public Announcements Multiagent epistemic logic with public announcements has been proposed and axiomatized by Plaza (1989), and, independently, by Gerbrandy and Groeneveld (1997). In (Plaza 1989), public announcement is seen as a binary operation $+$, such that $\varphi + \psi$ is equivalent to $\langle \varphi \rangle \psi$. Following that, the logic of public announcements with common knowledge was axiomatized by Baltag et al. (1998) (in a paper also containing action model logic).

There are a fair number of precursors of public announcement logic. First, there is a prior and independent line on meta-level descriptions of epistemic change, in so-called interpreted systems, in what is known as the runs-and-systems approach. Many results that were later obtained in dynamic epistemic logic should be seen as special cases of a more general temporal epistemic framework developed in that community. Dynamic features are expressed with temporal modalities and not with dynamic modalities. A comprehensive overview is beyond the scope of these notes. We refer to the very readable "Reasoning about Knowledge" by Fagin et al. (1995), and for example to van der Meyden (1998).

Another prior line of research that we feel more comfortable to present in some detail (given the background of the authors) is dynamic modal approaches to semantics that are not necessarily dynamic epistemic. An approach roughly known as "dynamic semantics" or "update semantics" was pioneered by van Emde Boas et al. (1984), Landman (1986), Groeneveld (1995), and Veltman (1996). There are strong relations between that and more *PDL*-motivated work initiated by van Benthem (1989), and followed up by de Rijke (1994) and Jaspars (1994). A good background to read up on this is van Benthem (1996). All such approaches use dynamic modal operators for information change, but typically without epistemic modalities, not multiagent, and not with the "computable" change as in public announcements, where the description of the action allows to construct the new state of information. More motivated by the runs-and-systems approach mentioned above is van Linder et al. (1995). The *PDL*-related and interpreted system-related approaches assume (and do not construct) a transition relation between states.

An approach somewhat related to Gerbrandy (1999) is by Lomuscio and Ryan (1998). They do not define dynamic modal operators in the language, but they define epistemic state transformers that clearly correspond to the interpretation of such operators: $M * \varphi$ is the result of refining epistemic model

M with a formula φ, etc. Their semantics for updates is only an approximation of public announcement logic, as the operation is only defined for finite (approximations of) models.

Public announcement logic (and most other dynamic epistemic logics) is not a *normal* modal logic, as it is not closed under uniform substitution of propositional variables. For example, $[p]p$ is valid, but $[p \wedge \neg K_a p](p \wedge \neg K_a p)$ is not valid. That public announcement logic without common knowledge is compact and strongly complete was already shown by Baltag et al. (1999). Public announcement logic with common knowledge is neither strongly complete, nor compact (due to the infinitary character of the common knowledge operator), see again Baltag et al. (1999) or related (for the common knowledge aspect) standard references such as Halpern and Moses (1992).

We have seen that every formula in public announcement logic is equivalent to a formula in epistemic logic. The logics are equally expressive. But in public announcement logic you can express logical properties with fewer symbols (you can simply count the number of symbols in a formula as if it were a string of symbols): a public announcement formula can be exponentially shorter than the equivalent epistemic logical formula (there is a more precise way to say this, relating two infinite sequences of formulas in both logics). We say that public announcement is more *succinct* than epistemic logic. This matter is treated by Lutz (2006) for public announcement logic interpreted on models without special properties and by French et al. (2013) for the logic as presented here, for relations that are equivalence relations.

Unsuccessful Updates The history of unsuccessful updates starts with the Moore-sentences such as $p \wedge \neg K p$. The proper first reference on this is George E. Moore's "A reply to my critics," wherein he writes

> I went to the pictures last Tuesday, but I don't believe that I did is a perfectly absurd thing to say, although *what* is asserted is something which is perfectly possibly logically. (Moore 1942, p. 543)

The further development of this notion firstly puts Moore-sentences in a multi-agent perspective of announcements of the form "*p* is true and *you* don't believe that," and secondly in the dynamic perspective of the unsuccessful update that cannot be believed after being announced. An excellent list of references on the topic is found in (Hintikka 1962, p. 64).

The term "unsuccessful update" was coined by Gerbrandy (1999); see also Gerbrandy (2007). The word "unsuccessful" refers to the *postulate of success* in belief revision (Alchourrón et al. 1985) that requires new information to be believed in the resulting state of information. As we have seen, this is not desirable in dynamic epistemic logic. The definition of a successful formula φ as a formula for which $[\varphi]\varphi$ is valid is by van Ditmarsch and Kooi (2006). They also provide preliminary results on what formulas are successful. The full characterization of single-agent successful formulas is by Holliday and Icard (2010). The multiagent case is still open.

Action Models The action model framework has been developed by Baltag et al. (1998). The final form of their semantics is Baltag and Moss (2004). Their completeness proof has been simplified in van Ditmarsch et al. (2007). A more general setup for action model logic, with a completeness proof by way of PDL, is van Benthem et al. (2006). Alternative frameworks to model epistemic actions are by Gerbrandy (1999) and (restricted to $S5$ models) by van Ditmarsch (2000, 2002b). A version of the latter involving concurrency is van Ditmarsch et al. (2003).

Action model logic mixes syntax and semantics in a way that not everybody in the community is comfortable with (an action model is a nearly semantic object, with a domain and an accessibility relation; but it also features as the parameter of a dynamic modal operator, in the syntax), and alternative approaches to model epistemic actions keep popping up now and then. We refer to treatments involving action languages, such as Kooi (2003) and Aucher (2010), and a treatment that goes under the name of arrow updates (Kooi and Renne 2011). Such alternative approaches typically (to our knowledge) do not result in logics with other expressivity than the action model logic that permeates the community.

We have not been treating complexity issues (of satisfiability and of model checking) of dynamic epistemic logics systematically (for one thing, we would have to introduce and explain complexity classes first). For that, we refer to, for example, Lutz (2006) and Aucher and Schwarzentruber (2013). For the complexities of epistemic logics, see Halpern et al. (2004).

Belief Revision in Dynamic Epistemic Logic A link between belief revision and modal logic, i.e., explicit belief modalities and belief change modalities in the logical language, was made in a strand of research known as *dynamic doxastic logic*. This was proposed and investigated by Segerberg and collaborators in works such as Segerberg (1999); Lindström and Rabinowicz (1999); Segerberg (1998). These works are distinct from other approaches to belief revision in modal logics without dynamic modal operators, such as Friedman and Halpern (1994); Board (2004); Bonanno (2005) that also influenced the development of dynamic logics combining knowledge and belief change (Friedman and Halpern (1994) is in the tradition of temporal epistemic logic, that we do not do justice in this dynamic modally focused historical overview). In dynamic doxastic logics, belief operators are in the logical language, and belief revision operators are dynamic modalities. Higher-order belief change, i.e., to revise one's beliefs about one's own or other agents' beliefs and ignorance, are considered problematic in dynamic doxastic logic, see Lindström and Rabinowicz (1999).

Belief revision in dynamic epistemic logic was initiated in Aucher (2005); van Ditmarsch (2005); van Benthem (2007); Baltag and Smets (2008). From these, Aucher (2005); van Ditmarsch (2005) propose a treatment involving degrees of belief, more common in areas related to artificial intelligence; van Benthem (2007); Baltag and Smets (2008) propose conditional belief, a logically more suitable approach. Many more works on dynamic belief revision have appeared since, e.g. Girard (2008); Liu (2008); Dégremont (2011).

Beyond Dynamic Epistemic Logic   Recent overview works also treating new developments in dynamic epistemic logic are van Benthem (2011) and van Ditmarsch et al. (2015).

13
Answers

13.1 Answers to Puzzles from Chapter 1

Answer to Puzzle 1
The third and fourth announcements are now false.

* Anne: "I know your number."
* Bill: "I know your number."

Instead, Anne and Bill both have to say once more that they do not know the other's number.

* Anne: "I don't know your number."
* Bill: "I don't know your number."

After Anne has said for the second time that she does not know Bill's number, the number pairs $(1, 2)$ and $(2, 3)$ are eliminated. After Bill says it for the second time, the number pairs $(3, 2)$ and $(4, 3)$ are eliminated. After that, Anne and Bill can announce truthfully that they know the other's number. Altogether, we therefore get (starting in the initial situation):

* Anne: "I don't know your number."
* Bill: "I don't know your number."
* Anne: "I don't know your number."
* Bill: "I don't know your number."
* Anne: "I know your number."
* Bill: "I know your number."

Answer to Puzzle 2
You cannot see what is written on your own forehead. But you can see what is on the forehead of the person who you are talking to. In the original version of the riddle, you only know your own number, but not the other's number. In this version, you only know the other's number, but not your own number. Otherwise, there is no difference. After the announcements the remaining number pairs are now $(2, 1)$ and $(3, 2)$, not $(1, 2)$ and $(2, 3)$.

Answer to Puzzle 3

If the numbers are m apart, then there are not 2 infinite chains of number pairs, but $2m$ such chains. For $m = 2$ (the numbers are two apart) we get,

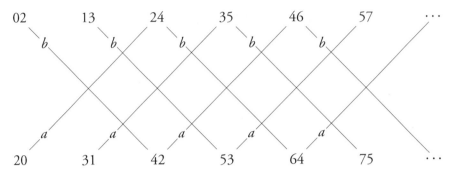

We now have four infinite chains with number pairs. After the two ignorance announcements

* Anne: "I don't know your number."
* Bill: "I don't know your number."

the remaining number pairs in the model where Anne now knows the numbers are $(2, 4)$, $(3, 5)$, $(4, 6)$, and $(5, 7)$. These are all in the upper part of the above picture. If Anne now says that she knows Bill's number, these four pairs are all that remains of the model, and Bill's subsequent announcement that he knows Anne's number does not change that. These four pairs do not have a property in common that would be "fun" for a problem solver to determine.

If the numbers are m apart, $2m$ number pairs remain after the two ignorance announcements.

Answer to Puzzle 4

If you see numbers that are two apart, then you know that your own number must be in between those. So in that case there is no uncertainty. If you see two consecutive numbers, then your number is one more than the largest or one less than the smallest. This also results in a model of infinite chains consisting of number triples connected by the indistinguishability relation. For example, let the numbers be 3, 4, and 5. There are 6 infinite chains of connected number triples. One of those is

$$012\text{---}a\text{---}312\text{---}b\text{---}342\text{---}c\text{---}345\text{---}a\text{---}645\text{---}b\text{---} \cdots$$

The other five infinite chains with possible number triples have as roots the triples 021, 201, 102, 120, and 210. If Anne says "I don't know my number," much uncertainty is eliminated all at once from the model. We get

$$012\text{---}a\text{---}312 \qquad\qquad 345\text{---}a\text{---}645 \qquad \cdots$$

The triple $(3, 4, 2)$ has been removed. Anne then sees 4 and 2 and therefore knows that she has 3. Infinitely, many such triples are removed from the chain in this way after Anne's announcement. Bill and Catherine are no longer uncertain after Anne's announcement. They know what the numbers are, whatever they are. Also, this is common knowledge. In the other five chains of number triples this story is repeated. Therefore, if Bill now says "I know my number," this is not informative. If Catherine were to say that, it is also not informative. And if Anne were to say "I don't know my number," then that is also not informative, because it is already common knowledge. In this version, we do not get more information after each knowledge or ignorance announcement. This version of the riddle is not much fun.

Answer to Puzzle 5

Anne and Bill have a natural number on their forehead of which the sum is 3 or 5. This comes with the following uncertainty:

$$(1, 4)-b-(1, 2)-a-(3, 2)-b-(3, 0)-a-(5, 0)$$

We can now have the same conversation as in the consecutive number riddle, in the "forehead" version where the agents do not know their own number, and this then results in the following information transitions. The final two announcements are not informative. In the Paterson et al. version, the conversation would have stopped after Anne's announcement "I know your number."

$$(1, 4)-b-(1, 2)-a-(3, 2)-b-(3, 0)-a-(5, 0)$$

* Anne: "I don't know my number."

$$(1, 2)-a-(3, 2)-b-(3, 0)-a-(5, 0)$$

* Bill: "I don't know my number."

$$(3, 2)-b-(3, 0)$$

* Anne: "I know my number."

$$(3, 2)-b-(3, 0)$$

* Bill: "I know my number."

$$(3, 2)-b-(3, 0)$$

13.2   Answers to Puzzles from Chapter 2

Answer to Puzzle 6
The prisoner can deduce that the hanging will take place on Friday. This is because the prisoner will know after night falls on Thursday, that the hanging is on Friday, and only then. In all other cases, he will remain uncertain when the hanging will be.

Answer to Puzzle 7
Rineke can conclude from overhearing the staffroom conversation that the exam will not be on Friday. But from overhearing the bikeshed conversation, she can further conclude that the exam will also not be on Thursday. As Friday had already been ruled out, the last possible day of the exam is now Thursday. So, in that case, when leaving school on Wednesday, she knows that the exam will be on Thursday, and that will then not be a surprise. Given that the teacher says that it remains a surprise, even given that Rineke knows that the exam cannot be on Friday, it therefore can also not be on Thursday. Just like before, Rineke cannot rule out any other day. By her last comment the teacher may of course have wasted the surprise for her pupil Rineke.

13.3   Answers to Puzzles from Chapter 3

Answer to Puzzle 15
Let us consider again the initial situation with all possible situations.

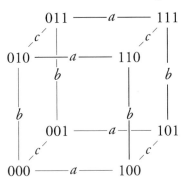

If father tells the three children that at least one of them is muddy, then, just like before, we can rule out 000. The children now say, simultaneously "I already knew." So they do not say "I already *know*," but they use the past tense instead. This refers to what is true and false in the initial situation pictured above (and that includes 000).

Consider the possibility 001. Caroline cannot distinguish this from 000. She sees two clean children. Therefore, she does not know that at least one child (herself) is muddy. Therefore she will not say "I already knew" after father says that at least one child is muddy. The same holds for Alice in 100 and Bob in 010. In the situation wherein 000 has already been removed, we can therefore, after the chorus "I already knew," also remove the possibilities 001, 010, and 100. The following picture results.

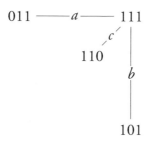

This figure also resulted from father's announcement "I will clap my hands. If you know that you're muddy, please step forward." Of course the three situations wherein a child can say "I already knew" are precisely those wherein it steps forward because it knows that it is muddy.

Because Alice, Bob, and Caroline are all muddy, father has to clap his hands twice after the "I already knew" response before they all step forward, instead of thrice in Puzzle 13.

Answer to Puzzle 16

1. Alice steps forward after the first clapping of hands, whoever may be muddy, and independently of whether she was initially muddy. It does not matter. She knows that she is clean, because father just cleaned her with a towel. If only Alice is muddy, then no one else will step forward at the second and the third clapping of hands. This may seem strange: if Alice had not been cleaned, she would have stepped forward anyway, because she would then have known that she is muddy. And then Bob and Caroline would have stepped forward the second time father clapped his hands, because they would have learnt that they were clean. Now, they remain uncertain: Alice steps forward because she has been cleaned, not because she sees that Bob and Caroline are clean. Therefore, Bob and Caroline remain standing the second time father claps his hands, and also the third time.

2. If only Alice and Bob are muddy, Alice steps forward at the first clapping of hands, and again nothing will happen the second and third time father claps his hands. This may seem strange, as Bob is muddy.

Bob sees that Alice is muddy, and knows from father's announcement that at least one child is muddy. He does not know whether he is muddy. When Alice has been cleaned he still does not know. Now when father claps his hands, Alice steps forward. In the previous version of the riddle Bob can deduce that Alice is only seeing clean children from that, and that he therefore must be clean. But now he cannot deduce that. Just as well, because he is muddy! Therefore, he does not step forward the second time. But therefore, he will also not step forward the third time. (The situation is similar to when only Alice is muddy.)

3. If only Bob and Caroline are muddy, they will step forward at the second clapping of hands. They can both see that Alice was already initially clean, and that father's action does not really *make* her clean. They are only uncertain about being muddy themselves. After the first clapping of hands only Alice steps forward, but not Bob and also not Caroline. From Caroline not stepping forward Bob learns that he is muddy, and from Bob not stepping forward Caroline learns that she is muddy. Therefore, they both step forward at the second clapping of hands.

4. When they are all muddy, again only Alice steps forward, and again nothing happens at the second and third clapping of hands. It is a bit harder to see why. Bob and Caroline learn nothing from only Alice stepping forward at the first hand clapping. Everyone knows that only Alice will step forward, because initially everybody sees two muddy children. At the second clapping of hands it is now no longer the case that, if only Alice and Bob had been muddy or only Alice and Caroline, these two would now have stepped forward. (See the second item, the situation for Alice and Caroline muddy is like that for Alice and Bob muddy.) For Bob and Caroline to step forward at the third clapping of hands, the (original) situations 110, 101, and 011 would now have had to be ruled out. But only 011 is ruled out (see previous item). Therefore Bob and Caroline remain uncertain whether they are muddy.

Figure 13.1, which we will not explain further, may clarify the answers of this puzzle.

Answer to Puzzle 17

Alice and Bob are muddy, and Alice will step forward at the first clapping of hands.

1. Bob concludes that he is clean. He believes that Alice stepped forward because she sees no muddy child, and had therefore concluded that she is muddy. But Bob is mistaken. He is muddy.

2. Caroline sees that Alice and Bob are muddy. She does not know whether she is muddy herself. She therefore does not know yet if Alice and Bob

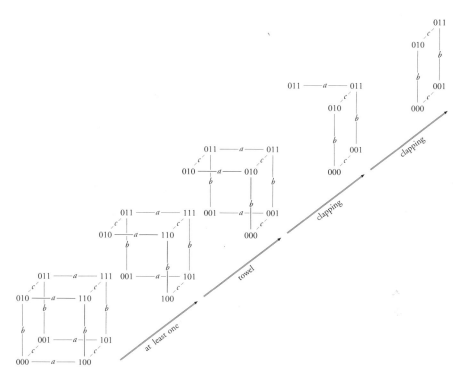

Fig. 13.1 Muddy children with cleaning

will step forward at the second or at the third clapping of hands. But she knows that clapping once is not enough: Alice cannot know whether she is muddy. Therefore, Caroline knows that Alice is *lying*. You are lying if you say that something is true when you know that it is false. Of course Alice does not actually *say* that she knows that she is muddy, but she *acts as if* she knows that she is muddy: her action can be identified with such a lie.

3. If Alice thinks after all, she would realize that stepping forward was lying. She does not know if Bob knows that she is lying, because she does not know whether she is muddy herself. But she knows that Caroline knows that she is lying, because Alice and Caroline both see that Bob is muddy.

If all three are muddy, only Alice steps forward, and Bob and Caroline learn that Alice is lying. If she had been thinking about what she was doing, Alice would also have deduced that Bob and Caroline knew that she was lying.

Answer to Puzzle 18

Prior to standing in line, the children have agreed that the hindmost child will say "white" if it sees an even number of white hats, and that it will say "black" if it sees an odd number of white hats. Suppose that the hindmost child, let us call her Alice, sees four black and five white hats in front of her. She does not know the color of her own hat. Following the agreed protocol, Alice says "black," as the number of white hats is odd. This is just as likely to be true as it is likely to be false: she cannot correctly guess the color of her own hat. But with this protocol, all other children will be able to correctly guess the color of their hat.

Suppose Bob is standing in front of Alice. If he sees five white hats in front of him, he knows that his hat is black and says "black." If he sees four white hats in front of him, his hat must have been one of the five white hats Alice was seeing, so he knows that his hat is white, and says "white."

All other children can reason similarly and thus correctly announce their hat's color.

Answer to Puzzle 19

Suppose *C* has a red stamp on her forehead. *A* can only know which color she does not have, if *B* and *C* both have a red stamp on their forehead. Therefore, when *A* says that she does not know that, *B* learns that he has no red stamp on his forehead. If *B* were then asked if he knows a color he does not have on his forehead, he would answer "I know that the stamp on my forehead is not red." But *B* says that he does not know a color that he definitely does not have. Therefore, *C* cannot have a red stamp on her forehead. An analogous argument can be made on the assumption that *C* has a yellow stamp on her forehead. If *C* does not have a red or a yellow stamp, she must have a green stamp on her forehead.

The informative consequences of the two subsequent answers, by *A* and *B*, are depicted below. In the last state of information, *C* can only have a green stamp. (But one cannot derive what the colors are of the stamps of *A* and *B*.) The nodes are named by triples that stand for the colors of the stamps of *A*, *B*, and *C*, respectively. For example, *rgy* stands for: *A* has a red stamp, *B* has a green stamp, and *C* has a yellow stamp. In the pictures, we use lower case letters *a*, *b*, *c* instead of upper case letters for the names of the agents.

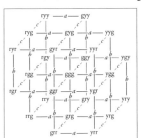 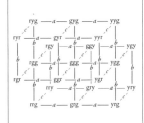 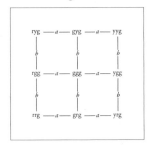

13.4   Answers to Puzzles from Chapter 4

Answer to Puzzle 22

The probability that the car is behind door 1 is 1/1000. Therefore, the probability that the car is behind any other door is, of course, 999/1000. Suppose you always switch doors. Similarly to before, the probability to get the car then becomes $\frac{1}{1000} \cdot 0 + \frac{999}{1000} \cdot 1 = \frac{999}{1000}$. After opening all those doors, it seems already much more likely that the car must then be behind door 899, which is indeed the case. But the analysis is the same as for the case of three doors.

Answer to Puzzle 23

Yes, it is rational to switch doors. It is now certain that the car is behind door 2. If there had been no car behind door 2, then the quizmaster would have opened that door instead of door 3. Because he opened door 3, the car must be behind door 2.

Answer to Puzzle 24

It is not rational to switch, but also not irrational. Opening door number 3 just rules out that the car is there, but is otherwise uninformative. Let us see why this is the case.

In this scenario, we can consider the door that the host opens to be a signal. So the strategies available are different than just always switching or never switching, because we can let the decision to switch or not switch depend on which door is opened. The arguments from the original scenario still hold; always switching or never switching still give the probability of winning the car to be $\frac{2}{3}$ and $\frac{1}{3}$, respectively.

What signal does the host give by opening door number 3? Given that the host wants to walk as much as possible and that door number 3 is the furthest he can walk, he opens door number 3 when the rules allow it. This makes opening door number 3 a lot less informative in this scenario.

When he opens door number 2, that must mean that the car is behind door number 3 (otherwise he would have opened door number 3). So we should always switch when he opens door number 2. And we are guaranteed to win the car in this case. Now, suppose we also switch when he opens door number 3. This means that we always switch, regardless of the door that the host opens. Since we know that always switching will win the car in two-thirds of the cases, and we are guaranteed to win when he opens door number 2 and the car is behind door number 3 (which occurs in just one third of the cases), it follows that by also switching when he opens door number 3, the chance of winning is only $\frac{1}{2}$ in that case in order to make the total chance of winning $\frac{2}{3}$. That is because $\frac{1}{3} \cdot 1 + \frac{2}{3} \cdot \frac{1}{2} = \frac{2}{3}$.

In other words, if the car is behind door 3 so that door 2 is opened (probability $\frac{1}{3}$), then the probability of winning by switching is 1, so that the probability of winning by not switching is 0; whereas, if the car is not behind door 3, so that door 1 is opened (probability $\frac{2}{3}$), then the probability of winning by switching is $\frac{1}{2}$, as computed above, so that the remaining probability of winning by not switching must also be $\frac{1}{2}$. Therefore, in that case, the probability to win by switching is equal to the probability to win by not switching.

Answer to Puzzle 25

It is not rational to switch, but also not irrational. The probability that the car is behind door 2 is $\frac{1}{2}$ and the probability that the car is behind door 1 is also $\frac{1}{2}$. In the original riddle, it was the host's knowledge that determined which door was opened. But now this is not the case. Door 1 or door 3 could have opened just as well due to technical error and the car might have been behind it.

This result is also easily simulated at home. Choose someone who does not know which door hides the car (for example, the candidate having to choose between doors!) and let that person open any door.

13.5 Answers to Puzzles from Chapter 5

Answer to Puzzle 26

Alice says: "I have one of 012, 034, 056, 135, 146, 236, 245."

Bob learns Alice's cards, because 3, 4, or 5 occur in each hand of cards except 012. We can repeat a similar computation for other possible hands of cards for Alice and for Bob: given a hand of cards for Alice, take any three out of the four remaining cards; then of those three cards will occur in one of the other seven hands. In other words, any hand of cards Bob may have will rule out all but one of the seven hands of the announcements. So Bob will always learn Alice's cards.

Cath learns none of Alice's cards and also none of Bob's cards, whatever the card deal wherein Alice can truthfully make her announcement, in other words: whatever the actual card of Cath had been.

Suppose Cath has card 0. The remaining possible hands for Alice are then: 135, 146, 236, 245. Card 1 occurs in hand 135. So Cath considers it possible that Alice has 1. Card 1 does not occur in 236. So Cath considers it possible that Bob has 1.

Suppose Cath has card 1. The remaining possible hands for Alice are then: 034, 056, 236, 245. Card 0 occurs in hand 034. So Cath considers it possible that Alice has 0. Card 0 does not occur in 236. So Cath considers it possible that Bob has card 0. Etc.

 . . . and after that Bob says "Cath has card 6."

The processing of Bob's announcement is the same as before. Schematically, the result of the first announcement is

012.345.6	012.346.5	012.356.4	012.456.3			
034.125.6	034.126.5			034.156.2	034.256.1	
		056.123.4	056.124.3	056.134.2	056.234.1	
135.024.6		135.026.4		135.046.2		135.246.0
	146.023.5		146.025.3	146.035.2		146.235.0
	236.014.5	236.015.4			236.045.1	236.145.0
245.013.6			245.016.3		245.036.1	245.136.0

and the result of the second announcement is

012.345.6
034.125.6

135.024.6

245.013.6

Answer to Puzzle 27

Alice announces the sum modulo 7 of her cards, after which Bob announces Cath's card.

For card deal 012.345.6 we therefore get

Alice announces "The sum of my cards is 3 modulo 7," after which Bob announces "Cath has card 6."

Apart from the actual hand 012, the other triples with sum 3 modulo 7 are: 046, 145, 136, 235. (For example, $2 + 3 + 5 = 10$ and 10 modulo 7 is 3, the number between 0 and 6 such that adding multiples of 7 makes it 10. Adding 7 just once is sufficient as $3 + 7 = 10$.) The further treatment of this solution is just as for the solution wherein the first announcement gives the alternative hands 012 034 056 135 246 (it is merely another execution of the same protocol). The "modulo sum" protocol always gives a solution, whatever Alice's hand of cards. These are the possibilities:

0	034, 025, 016, 124, 356
1	026, 035, 125, 134, 456
2	036, 045, 126, 135, 234
3	012, 046, 145, 136, 235
4	etc.
5	
6	

To realize that 012, 046, 145, 136, 235 and 012, 034, 056, 135, 246 are not so different, note that the permutation (1024563) transforms the former into the latter. (A permutation can be seen as a bijective function from a set to itself. The notation above is a shorthand for the function that maps 0123456 (in that order) to 1024563, i.e., a function f such that $f(0) = 0$, $f(1) = 0$, $f(2) = 2$, etc.)

Answer to Puzzle 28
Cath may now learn one of Alice's or Bob's cards, merely not their entire hand of cards. For example:

Alice says "My hand is one of 012, 034, and 056."

After this, it is common knowledge that Alice has card 0 and that Bob knows Alice's hand, so that Bob can again announce Cath's card.

Answer to Puzzle 29
Alice, Bob, and Cath hold, respectively, 4, 7, and 2 cards. The actual deal of cards is 0123.456789A.BC. The solution is known as the lines of a projective plane for $13 = 3^2 + 3 + 1$ points. (The seven hand solution for the Russian Cards problem is also known as the lines of a projective plane for $7 = 2^2 + 2 + 1$ points.) Observe that each pair of numbers (of which there are 78, the number of combinations of 2 out of 13) occurs once only in the answer. The answer is that Alice announces that her hand is one of the following 13 alternative 4-tuples.

0123	147A	248C	349B
0456	158B	259A	357C
0789	169C	267B	368A
0ABC			

We will not fully justify how we came up with this answer, but the following may help. Given the actual hand 0123, add *three* other hands wherein Alice holds 0, by using up all 9 remaining cards. Then, add three alternatives wherein she also holds 1, the first one by selecting one card from each of these three other hands also, the other two by making different choices from these three other hands (but such that no pair of cards occurs more than once in the 4-tuples so far). Then, do the same for 2 and for 3. Done.

Bob now knows Alice's hand of cards. He has the seven cards 456789A. We observe that one of these cards occurs in all hands except 0123. So he learns Alice's cards. This also holds for any other hand in Alice's announcement, for any hand Bob may then have. (The announcement is very symmetrical.) We can also reason by contradiction. Suppose Bob had not learnt the card deal from Alice's announcement. Then he would have considered at least $(6 - (2 + 2) =)$ two pairs of cards possible for Cath, and he would have considered at least two deals of cards possible. As there are six other cards given Bob's seven cards, there are then at least two cards that occur in both 4-tuples held by Alice in those two (or more) card deals. But no pair of cards occurs more than once in Alice's announcement. So, this is not possible.

Answer to Puzzle 30

Alice, Bob, and Cath hold respectively 2, 3, and 4 cards, and the eavesdropper Eve holds no card. Let us assume that the card deal is 01.234.5678.

Alice chooses any card not in her hands and announces that all her cards are among those three. For example,

Alice says "I hold two of the cards 0, 1, and 5."

The player holding that extra card, in this case, Cath, now knows the card deal. That player will now make the next announcement. At this stage, Bob, who does not hold any of the three cards, is uncertain between three card deals, namely 01.234.5678, 05.234.1678, and 15.234.0678. Alice of course is still just as uncertain as before her announcement. Cath now resolves the uncertainty of Alice and Bob with the following announcement.

Cath says "The card deal is one of 01.234.5678, 05.467.1238, and 15.678.0234."

The protocol underlying this announcement of three card deals requires that: (*i*) Alice holds 01, 05, and 15 in those deals (this guarantees that Alice learns the deal of cards); (*ii*) the deal wherein Alice holds 01 is the actual deal, and the unique one wherein Bob holds 234 (this guarantees that Bob learns the deal of cards); (*iii*) one of 0, 1, 5 occurs in Cath's hand in these three card deals, and no other card occurs in all of Bob's or all of Cath's hands in all three deals (this guarantees that Eve remains ignorant).

This then solves the riddle. Eve remains uncertain for any card, other than her own, what other player holds it: 0 may be held by Alice or by Cath, 1 may be held by Alice or Cath, 2 may be held by Bob or by Cath, etc. Eve knows that Alice does not have 2 (nor 3, nor . . .), but she still cannot point down the owner of the cards Alice does not have!

13.6   Answers to Puzzles from Chapter 6

Answer to Puzzle 31

If 0 is also permitted, then an agent knows its number if it sees a 0, in which case its own number is the other number it sees. For example, Alice sees 0 on Bob's forehead and 3 on Cath's forehead. She can then conclude that she also has a 3 on her forehead. Of course, Cath draws the same conclusion and only Bob, who is seeing two 3s, is uncertain about his number: he does not know if it is 0 or if it is 6; he cannot distinguish triple $(3, 0, 3)$ from triple $(3, 6, 3)$. In terms of the trees representing uncertainties, we simply enlarge the trunk at the root. To continue the example: we add a node 101 to the tree with root 121, and a branch labeled b linking 101 to 121: 101 is indistinguishable for Bob from 121.

The informative consequences of the combined three announcements on the three trees with roots $(1, 1, 0)$, $(1, 0, 1)$, and $(0, 1, 1)$ are in Figure 13.2. After those announcements, Alice knows her number in the situations $(2, 1, 1)$, $(5, 2, 3)$, $(3, 2, 1)$, and $(3, 1, 2)$. As Alice now says that her number is 50, the other numbers must then be either 20 and 30, or 25 and 25. You, as problem solver, cannot choose between the two! (Of course, Alice can, as she is seeing the other agents.) The problem cannot be solved.

Answer to Puzzle 32

Let there now be an upper limit for the numbers. For example, the numbers are at most 10. Now an agent will also know its number, if the sum of the numbers it is seeing is more than 10. For example, if the triple is $(2, 5, 7)$, then Alice knows that her number is 2. She sees 5 and 7, of which the sum is 12. But 12 is ruled out. Therefore, her number is 2: the difference of 5 and 7.

The following figure depicts the tree for upper limit 10, also allowing 0, and the result of three ignorance announcements in this structure. The number 10 is represented by the letter A (as in hexadecimal counting), to avoid ambiguity, as we do not write commas between the arguments of a number triple. We see that only the triples $(2, 1, 1)$ and $(2, 1, 3)$ remain. In either case, Alice would now know that her number is 2. Furthermore, Bob would now know that his number is 1. Only Cath remains uncertain.

The situation is also more complex in other ways if we work with upper limits, because it is no longer the case that a tree with multiple values for all

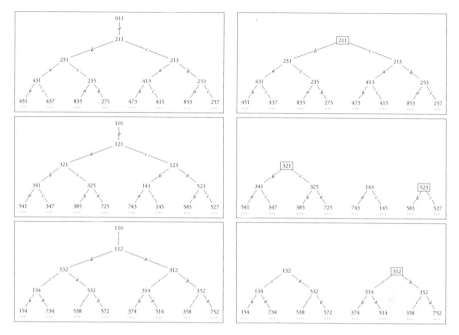

Fig. 13.2 Answer to Puzzle 31

arguments will be reduced the same way. We have to check all such "multiples of trees" separately. Already, nothing remains from the trees with root $(0, 2, 2)$ onward. For example, if the root is $(0, 5, 5)$, then only $(10, 5, 5)$ is considered possible. But as Bob says that he does not know the numbers, both these triples are ruled out. (In other words: Bob cannot make his announcement.) If the root is $(0, 6, 6)$, even the first ignorance announcement by Alice cannot be made. It is the same for all higher multiples. The trees with roots $(1, 0, 1)$ and $(1, 1, 0)$ are also entirely eliminated after the (maximum of) three ignorance announcements, and therefore also all their multiples. The only remaining triple is now $(0, 0, 0)$. But everybody knows their number already, so in that case no ignorance announcement can be truthfully made. So, $(0, 0, 0)$ is out as well.

Therefore, with upper boundary 10, Alice will always know her number after the three ignorance announcements have been processed, whatever the real situation is.

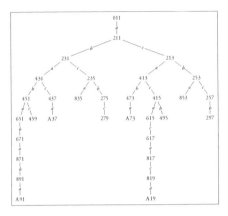
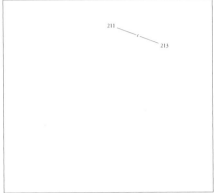

Some further calculations show us that Alice *always* knows her number after the three ignorance announcements if $8 \leq$ max ≤ 13, where max is the upper bound. If the numbers are at most 7, then not all three ignorance announcements can be made truthfully. But if the numbers are at most 14, then triples remain wherein Alice does not know her number.

13.7 Answers to Puzzles from Chapter 7

Answer to Puzzle 35
The third announcement in this version is the opposite of the third announcement in the original sum-and-product riddle. On the remaining ten sum lines we now keep the number pairs that were eliminated in the original version, and vice versa. So we keep the "open" number pairs wherein there is yet another product with a sum on one of the nine other sum lines. Again, this is informative for S. If the sum is 11, then there are three number pairs (namely $(2, 9)$, $(3, 8)$, and $(4, 7)$) wherein P knows the number pair and exactly one (namely $(5, 6)$) wherein he does not. For all other nine sums there always remains more than one number pair wherein P does not know it yet. (Please verify!) For example, for sum 17, these are (of course) all number pairs except $(4, 13)$. Schematically, we now get the following information transition.

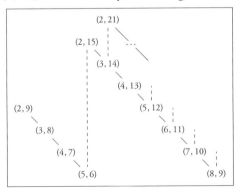
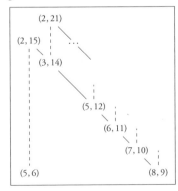

The fourth announcement is when S says "Now I know the numbers." The number pair must therefore be $(5, 6)$. The last announcement, wherein P also says "Now I know the numbers," is true but has no further informative content. This was already common knowledge between S and P.

Answer to Puzzle 36

The resulting model is

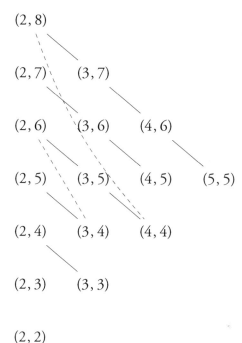

After P's announcement, four number pairs remain, not only $(2, 6)$ and $(3, 4)$, but also $(4, 4)$ and $(2, 8)$.

Answer to Puzzle 37

If x may be equal to y, then many number pairs with an even sum will be added. We now use the Goldbach conjecture again:

All even numbers larger than 2 are the sum of two primes.

This means that for all even sums, S does not know whether P knows the numbers, because they could have been prime.

We now add all pairs (x, x) to the system, for $x \leq 50$ (as $x + x \leq 100$).

⊛ If x is not prime, then (x, x) has the same sum as some other pair (z, w) with z and w prime (Goldbach), so it will be removed at the second announcement.

- If x is prime, then such a pair may make a difference, in principle, for the processing of the announcement "S knows that P does not know the numbers." We consider all different cases:
- The cases $x = 2$ and $x = 3$ were treated in Puzzle 36. This did not add uncertainty to the model. In both cases, P then already knows what the numbers are (so they will be removed at the second announcement).
- If $x > 3$, then there are always at least two prime number pairs with sum $2x$. (We do not give the proof.) The other pair, let us call it (z, w) again, already occurred in the model without equality. At the announcement "S knows that P does not know the numbers" all such pairs (x, x) and (z, w) will now be removed simultaneously, whereas before only (z, w) was removed.

Therefore, after the second announcement both models are again the same.

13.8   Answers to Puzzles from Chapter 8

Answer to Puzzle 38
You now certainly want to take the other envelope, because the argument for a higher expected value is now valid. This is because we now know how the probability distribution came about.

13.9   Answers to Puzzles from Chapter 9

Answer to Puzzle 39
We can adapt Protocol 4! The trick is that we have to compensate for the initial uncertainty about the state of the light in such a way that it does not matter for Anne's count, and that it applies to all non-counters. This is the solution:

Protocol 11 *The prisoners appoint one amongst them as the counter. All non-counting prisoners follow this protocol: the first two times they enter the room when the light is off, they turn it on; on all other occasions, they do nothing. The counter follows a different protocol. If the light is off when he enters the interrogation room, he does nothing. Whereas, if the light is on when he enters the interrogation room, he turns it off. When he is doing that for the 198th time, he also (truthfully) announces that everybody has been interrogated.*

For n prisoners the count is $2n - 2$. We explain why the protocol works for the case of three prisoners. Then, the counter counts to 4, i.e., four times turning off the light.

In case the light was initially on, that makes 1. Now suppose either Bob or Caroline has already turned on the light twice, suppose Bob, then that makes 2. And suppose Caroline then has only turned it on once, that makes 1. Together this now makes the required 4. But in case the light was initially off, Bob and Caroline have to turn it on twice, which makes again 4.

In this protocol the counter does not always have to wait until the last non-counter who has not yet turned on the light twice, has done so the second time.

Answer to Puzzle 40

Let us call a non-counter *lucky* if he or she announces (truthfully) that everybody has been interrogated before the counter does so, according to Protocol 5. There are two ways for Bob to get lucky:

1. If Bob's first interrogation is *before* Caroline's first interrogation (Bob turns on the light), then if Anne has subsequently been interrogated *before* Caroline (Anne turns off the light), Bob is interrogated again *before* Caroline; and after Caroline has subsequently been interrogated (for the first time; and, as the light is off, she now turns it on) Bob is interrogated *before* Anne. He will then announce that everybody has been interrogated.
2. If Caroline's first interrogation is *before* Bob's first interrogation (Caroline turns on the light), then Bob is interrogated *before* Anne. At some stage (after Anne has been interrogated and has turned off the light) Bob will now be interrogated again when the light is off. He will then turn on the light, as this is his role as a non-counter, and simultaneously he will announce that everybody has been interrogated.

The ways for Caroline to be lucky are similar. Just replace the roles of Bob and Caroline in the description.

Everywhere above where *before* is emphasized, this is a question of equal odds that one of two prisoners is interrogated before the other, on the assumption of random scheduling of interrogations. For example, for the first choice, the probability that Bob's interrogation is before Caroline's is $\frac{1}{2}$. As there are four such choice moments in the first item, the probability of that event happening is therefore $\frac{1}{2} \cdot \frac{1}{2} \cdot \frac{1}{2} \cdot \frac{1}{2} = \frac{1}{16}$. In the second item there are two choice moments, so this will carry a probability of $\frac{1}{4}$. This adds up to $\frac{5}{16}$. This is the probability that Bob is lucky. To this we have to add the equal probability that Caroline is lucky. Together this gives a probability of $\frac{5}{8}$ (i.e., 62.5%) that a non-counter is lucky; i.e., that Bob or Caroline announces before Anne that everybody has been interrogated.

For n larger than 3, it rapidly gets extremely rare that a non-counter is lucky. For 100 prisoners the probability that a non-counter is lucky is less than $5.63 \cdot 10^{-72}$.

Answer to Puzzle 41

The four prisoners are Anne, Bob, Caroline, and Dick, and the probabilities are $Pr(0) = Pr(1) = 1$, $Pr(2) = 0.5$, $Pr(3) = Pr(4) = 0$. We have enriched the execution sequence with annotation to explain how the protocol works. The lower index stands for the token (the number of points) of the prisoner whose name precedes that index, after performing the action according to Protocol 6. This is somewhat like the lower index in previous executions, where it stood for the number of *other* prisoners the counter had already counted. The difference is that the counter now also counts himself, so that the count is one higher at termination of the protocol. The upper index stands for the state of the light, as before. We recall that a prisoner's action depends on the sum of its token when entering the interrogation room and the state of the light. A prisoner's name is in bold if he collects a point, it is not in bold if he drops a point. As $Pr(2) = 0.5$, we can let a prisoner throw a coin and let heads stand for dropping a point and tails for collecting it.

$$^0\text{Anne}_0^1\text{Bob}_1^1\textbf{Caroline}_2^0\text{Dick}_0^1\textbf{Bob}_2^0\textbf{Caroline}_2^0\text{Caroline}_1^1\textbf{Bob}_3^0\text{Caroline}_0^1\textbf{Bob}_4^0$$

Anne gets there first, and turns on the light (= drops her point), then Bob comes in, flips a coin, heads, so does *not* turn off the light (= does *not* collect a point), then Caroline comes in, flips a coin, tails, so does turn off the light, then Dick, light on, then Bob again, who turns the light off this time and now has (again) 2 points. Then, first Caroline does not drop her point (tails/collect), but in her next interrogation she does and turns on the light (heads/drop), and this is subsequently collected by Bob. Crucially, at this point, Bob is designated as the "counter": as $Pr(3) = Pr(4) = 0$, Bob will never drop a point from here on but only collect them, until termination of the protocol. Anne and Dick already play no role anymore: once you have dropped your single point it does not matter whether the light is on or off at any subsequent interrogation, as $Pr(0) = Pr(1) = 1$, and, as already mentioned, dropping a point if you do not carry one, means doing nothing. The protocol now terminates by Caroline dropping another point and Bob collecting that final point.

It is important to realize that we must not have that $Pr(2) = 0$, because then a situation can be reached where two players "stick to their points" so that the protocol will never terminate. This situation occurs in the above sequence after the fifth interrogation. At that stage, Bob and Caroline both have a token of 2 points. We also cannot have $Pr(2) = 1$, because then no prisoner will ever get more than two points, and the protocol will also not terminate. Probability plays an essential role in this protocol.

Answer to Puzzle 42
First, a non-counter has to be interrogated, and has to turn on the light. The probability for this to happen is 99/100. Then, the counter has to be interrogated, to turn off the light again. The probability for that to happen is 1/100. After that, a non-counter who has not yet turned on the light, has to be interrogated and turn on the light. The probability for that is 98/100. After that comes the counter again, with always the same probability 1/100. And so on. If the probability for a daily event is p, then the expectation in number of days for that event to happen is $1/p$. Therefore, the expectation in number of days for the first non-counter to be interrogated is 100/99, which is just about a single day, the expectation for the counter to be interrogated after that is 100/1, 100 days, etc. The average number of days before the counter can declare that everybody has been investigated is therefore:

$$\frac{100}{99} + 100 + \frac{100}{98} + 100 + \ldots + \frac{100}{2} + 100 + \frac{100}{1} + 100.$$

The subsequence $\frac{100}{99} + \frac{100}{98} + \ldots + \frac{100}{2} + \frac{100}{1}$ can be rounded off to 518 days, and for the remainder we get 99 times 100 days, which is 9900 days. The sum of both figures is 10,418 days, which is about 28.5 years. That is a long time to wait to (maybe) go free if you are in prison.

13.10 Answers to Puzzles from Chapter 10

Answer to Puzzle 43
The maximum number of calls to distribute all secrets is the number of ways to choose two members out of a set of n elements: $\binom{n}{2} = \frac{n \cdot (n-1)}{2}$. This is, therefore, also the maximum number of different calls between n friends. For six friends a, b, c, d, e, f the following calls can be made such that in every call, at least one friend learns at least one secret—for convenience we generate the execution sequence in lexicographic order again.

ab; ac; ad; ae; af; bc; bd; be; bf; cd; ce; cf; de; df; ef

For four friends we get

ab; ac; ad; bc; bd; cd

Let us be explicit and give the detailed distribution of secrets for four friends:

	a	b	c	d
	A	B	C	D
ab	AB	AB	C	D
ac	ABC	AB	ABC	CD
ad	$ABCD$	AB	ABC	$ABCD$
bc	$ABCD$	ABC	ABC	$ABCD$
bd	$ABCD$	$ABCD$	ABC	$ABCD$
cd	$ABCD$	$ABCD$	$ABCD$	$ABCD$

Answer to Puzzle 44

If there are three friends, the expected number of calls in the Learn New Secrets protocol is 3. This is easy, because all executions of Learn New Secrets for three agents have length 3.

If there are four friends, the expected number of calls in the Learn New Secrets protocol is larger than 5. This we can see by comparing the number of executions of length 4 with the number of executions of length 6. (The number of executions of length 5, of which there are more, is somewhat harder to compute. But to get the answer, we do not need to do it.) This is because there are many more executions of length 6 than executions of length 4. Without loss of generality, assume that the first call is ab.

The typical execution sequence of length 4 is ab; cd; ac; bd. Instead of cd, the second call can (only) also have been dc (one out of 2). For the third call, either of the first callers calls either of the second callers, so the other options are $ca, ad, da, bc, cb, bd, db$ (one out of 8). Then, the last call is between the friends not making the third call, so the only other alternative is db (one of out 2). There are therefore 32 executions with first call ab.

The execution we used to prove that 6 is the maximum length, is ab; ac; ad; bc; bd; cd. (This is not the only type of execution of length 6, another one is ab; ac; bc; ad; bd; cd. But we do not need this.) For the second call there are 8 options: $ac, ca, ad, da, bc, cb, bd, db$. For the third call there are only 2 options, as the unique friend involved in both first two calls now calls the unique friend not yet involved in a call, or vice versa. The fourth call can only be from b to c, as c already knows B. In the fifth and sixth call we can reverse the charges of the call: two possibilities each. (For example, given the third call ad, the fourth call is between the agents not involved in the third call: bc or cb; etc.) This already makes 64 executions of length 6 starting with ab, more than the altogether 32 executions of length 4 starting with ab. And there are even more such executions of length 6, as we already argued.

Therefore, no matter how many length 5 executions there are, the average execution length of the Learn New Secrets protocol must be strictly larger than 5.

Answer to Puzzle 45

The solution sequence is *ab*; *cd*; *ac*; *bd*. Amal and Chandra are not involved in the last call. They do not know after the fourth call that everybody knows all secrets. After the three calls *ab*; *cd*; *ac*, the following calls are possible according to the protocol: *bd*, *bc*, *db*, and *da*. We can identify calls *bd* and *db* as they have the same informative effect. Amal and Chandra already know all secrets, so will not initiate a call. Only Bharat and Devi can initiate calls. They both do not know two (different) secrets yet. Given these four possibilities, Amal also considers it possible that the fourth call was *bc*, and Chandra also considers it possible that the fourth call was (*da*, i.e.,) *ad*. If the fourth call had been *bc*, after that call Amal, Bharat, and Chandra know all secrets, so the next call would have been initiated by Devi. She can now call either of these three and then everybody knows all secrets. If the fourth call had been *ad*, then a fifth call by Bharat would have been necessary to terminate the protocol. Either way, we will not reach the maximum of six calls. Amal and Chandra clearly know this. But Bharat and Devi can also come to this conclusion, because by a similar argument they can conclude that the third call must have been between Amal and Chandra.

It is a bit unclear if common knowledge is now already achieved. But (on the assumption that the protocol is common knowledge) after another 10 min, it certainly is.

Answer to Puzzle 46

Consider the sequence *ab*; *cd*; *ab*; *cd*; ... consisting of an infinite alternation of calls *ab* and calls *cd*. Just as the third call from *a* to *b* is justified because *a* considers it possible that the second call may have involved *b*, the fourth call from *c* to *d* is justified because *c* considers it possible that the third call may have involved *d*, and so on ... .

Answer to Puzzle 47

Let there be $n = 2^m$ friends. Let the n friends be named $1, \ldots, n$. We count modulo 2^m. The first round consists of 2^{m-1} parallel calls between two friends: for $i = 1$ to $i = 2^{m-1}$, all friends $2i - 1$ (simultaneously) call their neighbor $2i$ (i.e., for future convenience, $2i + 2^1 - 2$). The second round also consists of 2^{m-1} parallel calls, but now between friends that were not paired in the first round. A way to implement this is for all friends $2i - 1$ (simultaneously) to call friends $2i + 2$, i.e., $2i + 2^2 - 2$. (And nobody will find the line engaged!) We continue to do so m times altogether, namely until in the mth round all $2i - 1$ (simultaneously) call $2i + 2^m - 2$. For example, for eight friends a, b, c, d, e, f, g, h (i.e., $1, 2, \ldots, 8$) the three rounds are $\{ab, cd, ef, gh\}$; $\{ac, bd, eg, fh\}$; $\{ae, bf, cg, dh\}$. Let us be explicit again.

	a	b	c	d	e	f	g	h
	A	B	C	D	E	F	G	H
i	AB	AB	CD	CD	EF	EF	GH	GH
ii	$ABCD$	$ABCD$	$ABCD$	$ABCD$	$EFGH$	$EFGH$	$EFGH$	$EFGH$
iii	$ABCDEFGH$	$\ldots$	$\ldots$	$\ldots$	$\ldots$	$\ldots$	$\ldots$	$\ldots$

Answer to Puzzle 48
Let there be five friends. A four round parallel call sequence is: $\{ab, cd\}$; $\{ac, be\}$; $\{ae, bc\}$; $\{ad\}$. One can easily verify from the table below that less than four is indeed impossible.

	a	b	c	d	e
	A	B	C	D	E
$\{ab, cd\}$	AB	AB	CD	CD	E
$\{ac, be\}$	$ABCD$	ABE	$ABCD$	CD	ABE
$\{ae, bc\}$	$ABCDE$	$ABCDE$	$ABCDE$	CD	$ABCDE$
$\{ad\}$	$ABCDE$	$ABCDE$	$ABCDE$	$ABCDE$	$ABCDE$

Another configuration for the first two rounds starts with $\{ab, cd\}$; $\{ac, bd\}$; $\ldots$. But then we need three more rounds, and therefore five in total. A minimal completion of that is $\{ab, cd\}$; $\{ac, bd\}$; $\{ae\}$; $\{ab, ce\}$; $\{de\}$. Note that in the third round, there is nothing else to do but to make the single call between e and any other friend, as at this stage a, b, c, d already know all secrets except that of e, so there is no point anymore for them to call each other.

13.11   Answers to Puzzles from Chapter 11

Answer to Puzzle 49
Alice says "I do not have card 2," after which Bob says that he has won. As a result of Alice's announcement, the card deals 201 and 210 are ruled out. The information transition resulting from Alice's announcement is below.

Alice does not have card 2
$\Rightarrow$

Now, Bob announces that he knows the card deal. This is true if the card deal is 012 or 102, and false if the card deal is 021 and 120, because in the latter two cases Bob is uncertain between 021 and 120. Bob's anouncement therefore results in the elimination of 021 and 120. The transition is as follows:

Alice does not have card 2
$\Rightarrow$

In the resulting state of the game, Cath still does not know the card deal (she remains uncertain between 012 and 102), but Alice learns the card deal from Bob's announcement that he knows the card deal: she is therefore able to rule out 021.

Answer to Puzzle 50
Alice, Bob, and Cath each hold two cards from a pack of six for three suits, Wheat, Flax, and Rye. The actual card deal is *wx.wy.xy*. There are six card deals wherein all three players hold two cards of a different suit. All are relevant in order to determine what players know about each other. A model for this information is as follows (where *wx.wy.xy* is the actual card deal).

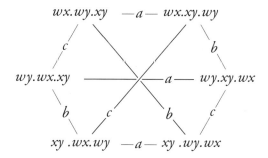

References

Albers, C. J., B. P. Kooi, and W. Schaafsma (2005). Trying to resolve the two-envelope problem. *Synthese 145*(1), 89–109.

Albert, M., R. Aldred, M. Atkinson, H. van Ditmarsch, and C. Handley (2005). Safe communication for card players by combinatorial designs for two-step protocols. *Australasian Journal of Combinatorics 33*, 33–46.

Alchourrón, C., P. Gärdenfors, and D. Makinson (1985). On the logic of theory change: Partial meet contraction and revision functions. *Journal of Symbolic Logic 50*, 510–530.

Attamah, M., H. van Ditmarsch, D. Grossi, and W. van der Hoek (2014). Knowledge and gossip. In *Proc. of 21st ECAI*, pp. 21–26. IOS Press.

Aucher, G. (2005). A combined system for update logic and belief revision. In *Proc. of 7th PRIMA*, pp. 1–17. Springer. LNAI 3371.

Aucher, G. (2010). Characterizing updates in dynamic epistemic logic. In *Proceedings of Twelfth KR*. AAAI Press.

Aucher, G. and F. Schwarzentruber (2013). On the complexity of dynamic epistemic logic. In *Proc. of 14th TARK*.

Aumann, R. (1976). Agreeing to disagree. *Annals of Statistics 4*(6), 1236–1239.

Baltag, A. and L. Moss (2004). Logics for epistemic programs. *Synthese 139*, 165–224.

Baltag, A. and S. Smets (2008). A qualitative theory of dynamic interactive belief revision. In *Proc. of 7th LOFT*, Texts in Logic and Games 3, pp. 13–60. Amsterdam University Press.

Baltag, A., L. Moss, and S. Solecki (1998). The logic of public announcements, common knowledge, and private suspicions. In *Proc. of 7th TARK*, pp. 43–56. Morgan Kaufmann.

Baltag, A., L. Moss, and S. Solecki (1999). The logic of public announcements, common knowledge, and private suspicions. Technical report, Centrum voor Wiskunde en Informatica, Amsterdam. CWI Report SEN-R9922.

Barwise, J. (1981). Scenes and other situations. *Journal of Philosophy 78*(7), 369–397.

Board, O. (2004). Dynamic interactive epistemology. *Games and Economic Behaviour 49*, 49–80.

Bonanno, G. (2005). A simple modal logic for belief revision. *Synthese 147*(2), 193–228.

Born, A., C. Hurkens, and G. Woeginger (2006). The Freudenthal problem and its ramifications: Part (I). *Bulletin of the EATCS 90*, 175–191.

Born, A., C. Hurkens, and G. Woeginger (2007). The Freudenthal problem and its ramifications: Part (II). *Bulletin of the EATCS 91*, 189–204.

Born, A., C. Hurkens, and G. Woeginger (2008). The Freudenthal problem and its ramifications: Part (III). *Bulletin of the EATCS 95*, 201–219.

Chow, T. (1998). The surprise examination or unexpected hanging paradox. *The American Mathematical Monthly 105*(1), 41–51.

Conway, J., M. Paterson, and U. Moscow (1977). A headache-causing problem. In J. Lenstra (Ed.), *Een pak met een korte broek—Papers presented to H.W. Lenstra, jr., on the occasion of the publication of his 'Euclidische Getallenlichamen'*, Amsterdam. Private publication.

Cordón-Franco, A., H. van Ditmarsch, D. Fernández-Duque, J. Joosten, and F. Soler-Toscano (2012). A secure additive protocol for card players. *Australasian Journal of Combinatorics 54*, 163–175.

de Rijke, M. (1994). Meeting some neighbours. In J. van Eijck and A. Visser (Eds.), *Logic and information flow*, Cambridge MA, pp. 170–195. MIT Press.

Dégremont, C. (2011). *The Temporal Mind. Observations on the logic of belief change in interactive systems*. Ph.D. thesis, University of Amsterdam. ILLC Dissertation Series DS-2010-03.

Dehaye, P., D. Ford, and H. Segerman (2003). One hundred prisoners and a lightbulb. *Mathematical Intelligencer 25*(4), 53–61.

Dixon, C. (2006). Using temporal logics of knowledge for specification and verification–a case study. *Journal of Applied Logic 4*(1), 50–78.

Fagin, R., J. Halpern, Y. Moses, and M. Vardi (1995). *Reasoning about Knowledge*. Cambridge MA: MIT Press.

Fernández-Duque, D. and V. Goranko (2014). Secure aggregation of distributed information. Online at http://arxiv.org/abs/1407.7582.

Fischer, M. and R. Wright (1992). Multiparty secret key exchange using a random deal of cards. In *Proc. of 11th CRYPTO*, pp. 141–155. Springer.

French, T., W. van der Hoek, P. Iliev, and B. Kooi (2013). On the succinctness of some modal logics. *Artificial Intelligence 197*, 56–85.

Freudenthal, H. (1969). Formulation of the sum-and-product problem. *Nieuw Archief voor Wiskunde 3(17)*, 152.

Freudenthal, H. (1970). Solution of the sum-and-product problem. *Nieuw Archief voor Wiskunde 3(18)*, 102–106.

Friedell, M. (1969). On the structure of shared awareness. *Behavioral Science 14*, 28–39.

Friedman, N. and J. Halpern (1994). A knowledge-based framework for belief change - part i: Foundations. In *Proc. of 5th TARK*, pp. 44–64. Morgan Kaufmann.

Gamow, G. and M. Stern (1958). *Puzzle-Math*. London: Macmillan.

Gardner, M. (1977). The "jump proof" and its similarity to the toppling of a row of dominoes. *Scientific American 236*, 128, 131–132, 134–135.

Gardner, M. (1979). Mathematical games. *Scientific American 241* (December), 20–24. Also addressed in the March (page 24) and May (pages 20–21) issues of volume 242, 1980.

Gardner, M. (1982). *aha! Gotcha: paradoxes to puzzle and delight*. New York: W.H. Freeman and Company.

Gerbrandy, J. (1999). *Bisimulations on Planet Kripke*. Ph. D. thesis, University of Amsterdam. ILLC Dissertation Series DS-1999-01.

Gerbrandy, J. (2007). The surprise examination. *Synthese 155(1)*, 21–33.

Gerbrandy, J. and W. Groeneveld (1997). Reasoning about information change. *Journal of Logic, Language, and Information 6*, 147–169.

Girard, P. (2008). *Modal logic for belief and preference change*. Ph. D. thesis, Stanford University. ILLC Dissertation Series DS-2008-04.

Grimm, J. and W. Grimm (1814). *Kinder- und Hausmärchen*. Reimer. Volume 1 (1812) and Volume 2 (1814).

Groeneveld, W. (1995). *Logical investigations into dynamic semantics*. Ph. D. thesis, University of Amsterdam. ILLC Dissertation Series DS-1995-18.

Halpern, J. and Y. Moses (1992). A guide to completeness and complexity for modal logics of knowledge and belief. *Artificial Intelligence 54*, 319–379.

Halpern, J., R. van der Meyden, and M. Vardi (2004). Complete axiomatizations for reasoning about knowledge and time. *SIAM Journal on Computing 33(3)*, 674–703.

Hardy, G. (1940). *A Mathematician's Apology*. Cambridge University Press.

Hedetniemi, S., S. Hedetniemi, and A. Liestman (1988). A survey of gossiping and broadcasting in communication networks. *Networks 18*, 319–349.

Hintikka, J. (1962). *Knowledge and Belief*. Ithaca, NY: Cornell University Press.

Holliday, W. and T. Icard (2010). Moorean phenomena in epistemic logic. In L. Beklemishev, V. Goranko, and V. Shehtman (Eds.), *Advances in Modal Logic 8*, pp. 178–199. College Publications.

Hurkens, C. (2000). Spreading gossip efficiently. *Nieuw Archief voor Wiskunde 5/1(2)*, 208–210.

Isaacs, I. (1995). The impossible problem revisited again. *The Mathematical Intelligencer 17(4)*, 4–6.

Jaspars, J. (1994). *Calculi for Constructive Communication*. Ph.D. thesis, University of Tilburg. ILLC Dissertation Series DS-1994-4, ITK Dissertation Series 1994-1.

Kirkman, T. (1847). On a problem in combinations. *Camb. and Dublin Math. J. 2*, 191–204.

Knödel, W. (1975). New gossips and telephones. *Discrete Mathematics 13*, 95.

Kooi, B. (2003). *Knowledge, Chance, and Change*. Ph.D. thesis, University of Groningen. ILLC Dissertation Series DS-2003-01.

Kooi, B. and B. Renne (2011). Arrow update logic. *Review of Symbolic Logic 4(4)*, 536–559.

Kooi, B. and J. van Benthem (2004). Reduction axioms for epistemic actions. In R. Schmidt, I. Pratt-Hartmann, M. Reynolds, and H. Wansing (Eds.), *Preliminary Proceedings of AiML-2004*, University of Manchester, pp. 197–211.

Kraitchik, M. (1943). *Mathematical Recreations*. London: George Allen & Unwin, Ltd.

Kvanvig, J. (1998). Paradoxes, epistemic. In E. Craig (Ed.), *Routledge Encyclopedia of Philosophy*, Volume 7, pp. 211–214. London: Routledge.

Landman, F. (1986). *Towards a Theory of Information*. Ph. D. thesis, University of Amsterdam.

Lewis, D. (1969). *Convention, a Philosophical Study*. Cambridge (MA): Harvard University Press.

Lindström, S. and W. Rabinowicz (1999). DDL unlimited: dynamic doxastic logic for introspective agents. *Erkenntnis 50*, 353–385.

Littlewood, J. (1953). *A Mathematician's Miscellany*. Methuen and Company.

Liu, A. (2004). Problem section: Problem 182. *Math Horizons 11*, 324.

Liu, F. (2008). *Changing for the Better: Preference Dynamics and Agent Diversity*. Ph.D. thesis, University of Amsterdam. ILLC Dissertation Series DS-2008-02.

Lomuscio, A. and M. Ryan (1998). An algorithmic approach to knowledge evolution. *Artificial Intelligence for Engineering Design, Analysis and Manufacturing 13(2)*, 119–132.

Lutz, C. (2006). Complexity and succinctness of public announcement logic. In *Proc. of the 5th AAMAS*, pp. 137–144.

Makarychev, K. and Y. Makarychev (2001). The importance of being formal. *Mathematical Intelligencer 23(1)*, 41–42.

McCarthy, J. (1990). Formalization of two puzzles involving knowledge. In V. Lifschitz (Ed.), *Formalizing Common Sense : Papers by John McCarthy*, Ablex Series in Artificial Intelligence. Norwood, N.J.: Ablex Publishing Corporation. original manuscript dated 1978–1981.

Meyer, J.-J. and W. van der Hoek (1995). *Epistemic Logic for AI and Computer Science*. Cambridge University Press. Cambridge Tracts in Theoretical Computer Science 41.

Moore, G. (1942). A reply to my critics. In P. Schilpp (Ed.), *The Philosophy of G.E. Moore*, pp. 535–677. Evanston IL: Northwestern University. The Library of Living Philosophers (volume 4).

Moses, Y., D. Dolev, and J. Halpern (1986). Cheating husbands and other stories: a case study in knowledge, action, and communication. *Distributed Computing 1(3)*, 167–176.

Mosteller, F. (1965). *Fifty Challenging Problems in Probability with Solutions*. Reading, Massachusetts: Addison-Wesley.

Nalebuff, B. (1989). The other person's envelope is always greener. *Journal of Economic Perspectives 3(1)*, 171–181.

O'Connor, D. (1948). Pragmatic paradoxes. *Mind 57*, 358–359.

Plaza, J. (1989). Logics of public communications. In *Proc. of the 4th ISMIS*, pp. 201–216. Oak Ridge National Laboratory.

Purvis, M., M. Nowostawski, S. Cranefield, and M. Oliveira (2004). Multi-agent interaction technology for peer-to-peer computing in electronic trading environments. In G. Moro, C. Sartori, and M. Singh (Eds.), *AP2PC*, pp. 150–161. Springer. LNCS 2872.

Quine, W. (1953). On a so-called paradox. *Mind 62*, 65–67.

Regis, G. (1832). *Meister Franz Rabelais der Arzeney Doctoren Gargantua und Pantagruel, usw.* Leipzig: Barth.

Sallows, L. (1995). The impossible problem. *The Mathematical Intelligencer 17(1)*, 27–33.

Scriven, M. (1951). Paradoxical announcements. *Mind 60*, 403–407.

Segerberg, K. (1998). Irrevocable belief revision in dynamic doxastic logic. *Notre Dame Journal of Formal Logic 39(3)*, 287–306.

Segerberg, K. (1999). Two traditions in the logic of belief: bringing them together. In H. Ohlbach and U. Reyle (Eds.), *Logic, Language, and Reasoning*, Dordrecht, pp. 135–147. Kluwer Academic Publishers.

Selvin, S. (1975a). On the Monty Hall Problem. *The American Statistician 29*(3), 134.

Selvin, S. (1975b). A problem in probability. *The American Statistician 29*(1), 67.

Shaw, R. (1958). The paradox of the unexpected examination. *Mind 67*, 382–384.

Sietsma, F. (2012). *Logics of Communication and Knowledge*. Ph. D. thesis, University of Amsterdam. ILLC Dissertation Series DS-2012-11.

Smullyan, R. (1982). *Lady or the Tiger? And Other Logic Puzzles Including a Mathematical Novel That Features Godel's Great Discovery*. New York: Random House.

Sorensen, R. (1988). *Blindspots*. Oxford: Clarendon Press.

Swanson, C. and D. Stinson (2014). Combinatorial solutions providing improved security for the generalized Russian cards problem. *Designs, Codes and Cryptography 72(2)*, 345–367.

Tijdeman, R. (1971). On a telephone problem. *Nieuw Archief voor Wiskunde 3(19)*, 188–192.

van Benthem, J. (1989). Semantic parallels in natural language and computation. In *Logic Colloquium '87*, Amsterdam. North-Holland.

van Benthem, J. (1996). *Exploring logical dynamics*. CSLI Publications.

van Benthem, J. (2007). Dynamic logic of belief revision. *Journal of Applied Non-Classical Logics 17(2)*, 129–155.

van Benthem, J. (2011). *Logical Dynamics of Information and Interaction*. Cambridge University Press.

van Benthem, J., J. van Eijck, and B. Kooi (2006). Logics of communication and change. *Information and Computation 204(11)*, 1620–1662.

van Ditmarsch, H. (2000). *Knowledge games*. Ph. D. thesis, University of Groningen. ILLC Dissertation Series DS-2000-06.

van Ditmarsch, H. (2002a). The description of game actions in Cluedo. In L. Petrosian and V. Mazalov (Eds.), *Game Theory and Applications*, Volume 8, pp. 1–28. Nova Science Publishers.

van Ditmarsch, H. (2002b). Descriptions of game actions. *Journal of Logic, Language and Information 11*, 349–365.

van Ditmarsch, H. (2002c). Het zeven-kaartenprobleem. *Nieuw Archief voor Wiskunde 5/3(4)*, 326–332.

van Ditmarsch, H. (2003). The Russian cards problem. *Studia Logica 75*, 31–62.

van Ditmarsch, H. (2005). Prolegomena to dynamic logic for belief revision. *Synthese 147*, 229–275.

van Ditmarsch, H. (2006). The logic of Pit. *Synthese 149(2)*, 343–375.

van Ditmarsch, H. (2007). Honderd gevangenen en een gloeilamp. *Nieuwe Wiskrant 27(1)*, 15–18.

van Ditmarsch, H. and B. Kooi (2005). Een analyse van de hangman-paradox in dynamische epistemische logica. *Algemeen Nederlands Tijdschrift voor Wijsbegeerte 97*(1), 16–30.

van Ditmarsch, H. and B. Kooi (2006). The secret of my success. *Synthese 151*, 201–232.

van Ditmarsch, H. and J. Ruan (2007). Model checking logic puzzles. In *Quatrièmes Journées Francophones MFI*, Paris, pp. 139–150. Annales du Lamsade, Université Paris Dauphine.

van Ditmarsch, H., W. van der Hoek, and B. Kooi (2003). Concurrent dynamic epistemic logic. In V. Hendricks, K. Jørgensen, and S. Pedersen (Eds.), *Knowledge Contributors*, Dordrecht, pp. 45–82. Kluwer Academic Publishers. Synthese Library Volume 322.

van Ditmarsch, H., J. Ruan, and R. Verbrugge (2007). Sum and product in dynamic epistemic logic. *Journal of Logic and Computation 18(4)*, 563–588.

van Ditmarsch, H., W. van der Hoek, and B. Kooi (2007). *Dynamic Epistemic Logic*, Volume 337 of *Synthese Library*. Springer.

van Ditmarsch, H., J. van Eijck, and R. Verbrugge (2009). Publieke werken: Freudenthal's som-en-productraadsel. *Nieuw Archief voor Wiskunde 5/10(2)*, 126–131.

van Ditmarsch, H., J. van Eijck, and W. Wu (2010a). One hundred prisoners and a lightbulb—logic and computation. In F. Lin, U. Sattler, and M. Truszczynski (Eds.), *Proc. of KR 2010 Toronto*, pp. 90–100.

van Ditmarsch, H., J. van Eijck, and W. Wu (2010b). Verifying one hundred prisoners and a lightbulb. *Journal of Applied Non-Classical Logics 20(3)*, 173–191.

van Ditmarsch, H., J. Halpern, W. van der Hoek, and B. Kooi (Eds.) (2015). *Handbook of epistemic logic*. College Publications.

van Emde Boas, P., J. Groenendijk, and M. Stokhof (1984). The Conway paradox: Its solution in an epistemic framework. In *Truth, Interpretation and Information: Selected Papers from the Third Amsterdam Colloquium*, pp. 159–182. Dordrecht: Foris Publications.

van Linder, B., W. van der Hoek, and J.-J. Meyer (1995). Actions that make you change your mind. In A. Laux and H. Wansing (Eds.), *Knowledge and Belief in Philosophy and Artificial Intelligence*, Berlin, pp. 103–146. Akademie Verlag.

van der Meyden, R. (1998). Common knowledge and update in finite environments. *Information and Computation 140(2)*, 115–157.

van Tilburg, G. (1956). Doe wel en zie niet om. *Katholieke Illustratie 90(32)*, 47. Breinbrouwsel 137.

Veltman, F. (1996). Defaults in update semantics. *Journal of Philosophical Logic 25*, 221–261.

vos Savant, M. (1990). Ask Marilyn. *Parade Magazine (Sept 9)*, 15.

Weiss, P. (1952). The prediction paradox. *Mind 61*, 265–269.

Winkler, P. (2004). *Mathematical Puzzles: A Connoisseur's Collection*. AK Peters.

Zabell, S. L. (1988a). Loss and gain: the exchange paradox. In J. M. Bernardo, M. H. DeGroot, D. V. Lindley, and A. F. M. Smith (Eds.), *Bayesian Statistics 3*, Oxford, pp. 233–236. Clarendon Press.

Zabell, S. L. (1988b). Symmetry and its discontents. In B. Skyrms and W. Harper (Eds.), *Causation, Chance and Crecedence*. Kluwer Academic Publishers.

Printed in the United States
By Bookmasters